Olive Oil

By
A.K. Kiritsakis
School of Food Technology and Nutrition
Technological Educational Institution (TEI)
of Thessaloniki, Greece

American Oil Chemists' Society
Champaign, Illinois

AOCS Mission Statement

To be a forum for the exchange of ideas, information and experience among those with a professional interest in the science and technology of fats, oils and related substances in ways that promote personal excellence and provide for a high standard of quality.

Library of Congress Cataloging-in-Publication Data

Kiritsakis, A. K.
[Elaiolado. English]
Olive oil / by A.K. Kiritsakis.
p. cm.
Includes bibliographical references.
ISBN 0-935315-32-2
1. Olive oil. I. Title.
TP683.K5713 1991
664'.362—dc20 91-6710
CIP

Printed in the United States of America.

Dedicated to
my wife Eleftheria

Preface

The history of the olive is as old as man himself. This volume begins in ancient times and progresses to modern scientific and technological viewpoints. The first chapter mainly covers the historical uses of olives and olive oils. The rest of the chapters discuss the scientific aspects relative to the botanical properties, the chemistry of the plant, as well as the technological and analytical aspects. Each topic is well-covered, and the editor and the authors display a deep knowledge of the science involved that derives from solid, daily experience with the subject.

Not only is this a complete and excellent volume from a scientific viewpoint, but it also espouses Kiritsakis' deep love of the olive plant, which I find symbolic of a love of humanity and its long journey, as the olive branch is now a symbol of good health as well as a symbol of peace.

Enzo Fedeli

Foreword

Many recent scientific publications emphasize the positive health effects of olive oil. The positive effects of olive oil consumption are reflected in the good health and life expectancy of the older generations of Cretans, who seem to have preserved their traditional dietary patterns for over 4000 years up until the last few decades. The beneficial effects of olive oil consumption within the framework of a healthy lifestyle appears to be the major reason for the low incidence of coronary heart disease mortality among the Cretans.

Although genetic engineering has succeeded in producing other vegetable oils with a similar fatty acids composition, the uniqueness of olive oil is unquestionable. Dr. Kiritsakis' book on olive oil provides an opportunity for researchers to have factual and detailed information on the history of olive oil production, its properties, composition, and biological effects at their fingertips. Dr. Kiritsakis' work is greatly needed, and will fill a vacuum in international literature. A publication on olive oil where specific attention is given to its effects on serum lipoprotein levels in relation to cardiovascular diseases and cancer prevention is of enormous significance within the field of preventive medicine.

It is an honor to introduce Dr. Kiritsakis' book which, I am sure, will be a valuable reference text for nutritionists, dieticians, physicians, and all related health and agricultural scientists.

Anthony Kafatos, M.D.

Notes

The olive tree has been cultivated for thousands of years around the Mediterranean Sea. The fruit and oil of this tree have contributed significantly to the diet of the Mediterranean people and is connected to the civilizations that flourished there. The olive branch symbolized friendship and peace among nations and the winners of the ancient Olympic games were crowned with branches of olive trees.

Olive oil has been used as a fuel and as a medicine, but modern technology and medicine have found replacements for these uses. Olive oil, however, is still a basic staple in the diet of the Mediterranean people and is appreciated internationally for its fragrant and delicate flavor. Recent findings indicate that olive oil may contribute to the prevention and control of cardiovascular or other diseases.

The author felt a special preditection for the study of olive oil early in his academic career. He is a native of Crete, a Greek island where the olive tree has been grown intensively since antiquity. Olive oil was the subject of his M.Sc and Ph.D research in the USA after earning a Greek B.S degree in Agronomy. He later worked in a Greek Olive and Subtropical Crops Experimental Station in Crete.

In this book olive oil is discussed in all its aspects: its biosynthesis and chemistry, the harvesting of the olive fruit, the extraction, processing, storage and packing of olive oil, as well as the quality control, deterioration, adulteration, analysis and marketing of olive oil. A special chapter is devoted to the nutritional qualities of olive oil.

I would like to thank the many people who have helped prepare my manuscript. First I thank my wife Eleftheria for her help in the original translation. I am really indebted to my teacher at Michigan State University and dear friend P. Markakis who reviewed the book, and made pertinent suggestions. His great assistance is really appreciated. My sincere appreciation is also extended to my major professor Dr. L.R. Dugan for making very useful suggestions for this book.

I am also indebted to Mr. G. Platsis for his great help in the literature review, to Dr. R. Hernandez for helping with the Spanish references and to Mr. A. Bizopoulos who helped me use the computer.

Acknowledgments are extended to Mr. F. Luchetti, executive director of the International Olive Oil Council (IOOC), and to Mr. J. Petras, vice president of the Technological Educational Institution (TEI) of Thessaloniki, for encouraging the English version of the book. The author also expresses his appreciation to the Central Cooperative Union of Olive and Oil Producers (ELEOURGIKI) for their financial support of the Greek edition of this book, and to the American Oil Chemists' Society (AOCS) for making the English edition possible.

My deepest gratitude is expressed to my wife Eleftheria and to my children Kostas and Ioanna, for their patience and sacrifices made during the preparation of this book. They have truly shared in its gestation, which was much longer than expected.

I am grateful to Dr. A. Kafatos for his ready response to the request to contribute the chapter "Biological Effects of Olive Oil on Human Health". I am also indebted to my friend Mr. A. Assimeonoglou who prepared the drawings. Finally, thanks to Mrs. A. Platara for her typing assistance.

Apostolos (Paul) Kiritsakis
Thessaloniki, Greece
March, 1991

Acknowledgments

This book was originally published in Greek and it received an award by the Academy of Athens. It was translated into English with the help of Mrs. E. Kiritsakis and P. Markakis. The author expresses his sincere appreciation for their great help.

The financial contributions of the following companies, organizations and cooperative unions* for the English version of this book are also deeply appreciated.

- SEVITEL
- INTERNATIONAL OLIVE OIL COUNCIL (IOOC)
- GIURLANI USA
- ELEOURGIA MESSINIAS (EL.ME s.a.)
- UNION OF AGRICULTURAL COOPERATIVES OF KOLYMVARI — CHANIA
- SUN — M. MOTAKIS s.a., OLIVE OIL BUSINESS
- ELAIS — OLEAGINOUS PRODUCTS s.a.
- MINERVA SOC. OLEICOLE s.a.
- ARTEMIS Ltd.
- ESVITE
- ARCHO Bros. Co.
- ASTRON MESSINIAS
- OLICO s.a.
- LASAC s.a. — L.K. KOLIAS & Co., OLIVE OIL INDUSTRIES
- UNION OF AGRICULTURAL COOPERATIVES OF KISSAMOS — CHANIA
- H.A.P. s.a. (Hellenic Agricul. Products)
- ELEOURGIA PELOPONNISSOU — TH. MARAGOS, Inc.
- SEMOLEA
- MERCHANTS' ASSOCIATION OF NORTHERN GREECE
- CH.J. THEOCHARIS s.a. — AGRICULTURAL MACHINE INDUSTRY
- AROMA — J.A. KOSTAKIS
- PANTAZATOY Brothers
- P. DEMOULAS, Inc.
- KATRAKAZOY Brothers and Sons
- CH. MORENOS and Son s.a.
- ELEOURGIKI — CENTRAL COOPERATIVE UNION OF OLIVE AND OIL PRODUCERS
- SUNFOODS HELLAS Ltd.

*Names were listed in the order of financial contributions.

Contents

Chapter 1

History of the Olive Tree

Olive trees are among the oldest known cultivated trees in the world. The botanical progenitor of the olive tree is not accurately known, but it is thought to be the Oleaster *Olea sylvestris* which is still grown wild in North Africa, Portugal, southern France, Italy and by the Black and Caspian Seas (1). According to another theory, the olive tree is a descendant of *Olea chrysophylla*, which grew in Ethiopia, Kenya, Uganda and neighboring areas. These two species probably originated from another which covered much of the Sahara Desert prior to the glacier age (1).

Unquestionably, the cultivation of the olive tree commenced before the written word was invented (2). De Candolle, along with other historians, considers Syria and Asia Minor, where wild olive trees grow in abundance on mountainsides, as the origins of the cultivated species. Wild olive trees, however, grown in North Africa, Spain and Greece render uncertain the exact origin of the olive tree.

Others believe that the olive tree originated from Africa (Ethiopia, Egypt). This is where olive trees were first cultivated systematically, and from where they spread to Cyprus, Morocco, Algeria, Tunisia and elsewhere by the Phoenicians. The historian Theophrastos reported that the olive tree grew in southern Italy, Syria, Arabia, Egypt and elsewhere. The olive tree was cultivated in Egypt thousands of years ago. It was around 2000 B.C. though that olive orchards disappeared, either because they were destroyed for some unknown reason or because the interest of the people turned to other crops. Perhaps at that time, olive trees were brought to the southern coast of Crete. Others believe that the olive tree was originally spread in the Greek islands and in the mainland of Greece from North Syria. Genetic and archaeological studies indicate that the original center of olive cultivation was Palestine, Lebanon, Syria, Cyprus and Crete (3,4).

Olive tree cultivation spread from Greece or North Africa to Italy and to other Mediterranean countries around 600 B.C. (5,3). The olive tree was probably introduced to Spain by the Greeks, the Romans and the Arabs. That is indicated by the use in Spain of both Greco-roman and Arabic words associated with olives. The olive fruit, for instance, is called "aceituna" and the olive oil "aceite" which are Arabic words, while the tree is called "olive" which is Latin from the Greek "elea."

The olive tree was widely cultivated in southern Europe and this is why it is called *Olea europea*. About 98% of the existing olive trees today are grown around the Mediterranean Sea. The Jewish people knew about the olive tree thousands of years ago. In the Hebrew culture, the olive tree symbolizes peace and happiness (5). The Romans considered those who used animal fats instead of olive oil in their diet as barbarians. Rome extended olive cultivation to the entire empire under its occupation (6).

The olive culture played an enormous role in the early civilizations of Egypt and Greece (5). Athens was named in honor of the goddess Athena who brought the olive tree to the city. Plato's Olive Tree on Holly street (Iera odos) in Athens still exists. Solon had passed special laws for the olive tree. It is believed that the god of farming and animal raising, Aristeos, invented the cultivation of the olive tree and the olive oil mill.

The olive tree is of great historical importance. It played an important role in areas such as diet; religion; and the decoration of pottery, of walls, and of gold pieces of art. It also constituted the symbol for peace, wisdom and victory. The crowning wreaths for the winners of the Olympic games were made of a wild olive tree branch (Kotinos).

The therapeutic properties of olive oil were known to Hippocrates, the father of medicine. The consumption of a spoonful of oil was common practice for many people, and still is for some despite the advancement of pharmacology.

Pictures found in the palace of Knossos in Crete indicate that people consumed olives, and used olive oil for cooking and for fuel in lamps. Archeologists believe that the wealth of the Minoic Kingdom was related to the successful trade of Cretan olive oil (1,3). Huge clay containers were used for the storage of olive oil can be found even today in ancient Knossos and Phestos and in other places.

Spreading of the Olive Tree in the Western Hemisphere

The olive tree was not native to the New World. Spanish and Portugal settlers as well as Franciscan missionaries brought the olive tree to California and to other subtropical areas of the Western Hemisphere (7).

Several cultivars were brought to California from the Mediterranean countries between the years 1850 to 1900, and it was clear already by 1875 that olive tree cultivation would be an asset both for the farmers and the economy of California. In that year there were about 11,500

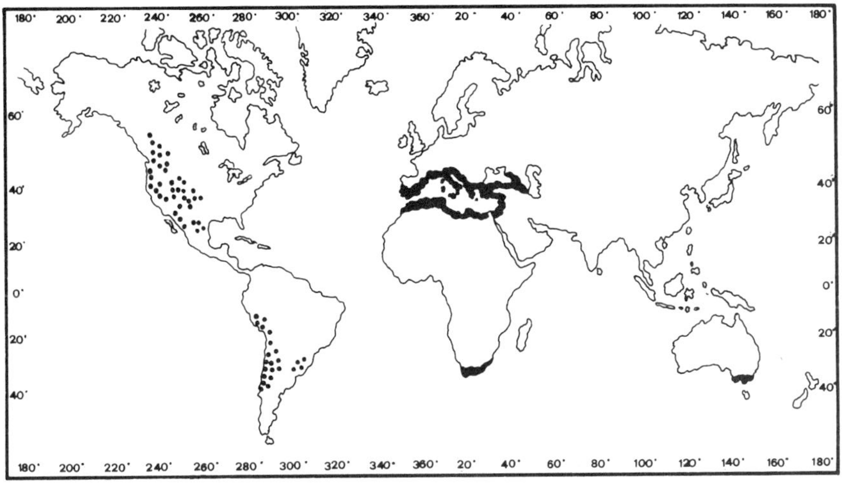

Fig. 1.1. Regions where the olive tree is cultivated.

olive trees; by 1910 their number grew to one million trees. People in California realized early that table olives gave more profit than olive oil. Thus many trees were grafted with table cultivars (8). It is the author's opinion that olive oil production should be increased in the United

TABLE 1.1

Geographical distribution of olive oil production.*

Europe 84%	Asia 5.5%	Africa 8.5%	Americas 2%
Spain	Turkey	Tunisia	Argentina
Italy	Syria	Algeria	U.S.A.
Greece	Lebanon	Morocco	Mexico
Portugal	Israel	Libya	Uraguay
France	Iraq	N. Africa	Peru
Soviet Union	Iran		
Yugoslavia	Japan		

*Reprinted with permission from Fedeli, E., Lipids of Olives. Prog. Chem. Fats and other Lipids 15:57 (1977).

States, since the Americans have started to realize the significance of this oil in their diet (See Chapter 17).

Italian immigrants brought the olive tree to Australia. The first trees were planted in Sidney around 1805. Efforts made to establish an olive industry in Australia, part of which is well suited to the cultivation of this tree, were not met with enthusiasm. As a result, most of the oil consumed was imported from Europe before World War II (9).

TABLE 1.2

World production and consumption of olive oil and table olives (Average 1978-1983 thousand metric tons).

Country	Number of olive trees (millions)	Olive oil		Table olives	
		Production	Consumption	Production	Consumption
Algeria	20.2	11.7	11.6	6.2	5.3
Argentina	5.0	11.1	2.0	31.0	11.9
Chile	0.8	0.2	0.3	3.2	3.3
Cyprus	2.6	1.7	1.9	3.7	4.0
Egypt	1.1	0.1	0.6	6.9	9.6
France	6.0	1.6	25.9	2.0	26.8
Greece	117.6	259.3	197.8	71.2	22.1
Israel	1.5	2.9	2.5	11.3	9.4
Italy	181.8	550.2	593.7	71.5	95.0
Jordan	2.7	8.2	9.9	7.8	8.0
Lebanon	6.0	5.2	6.9	5.6	6.0
Libya	8.0	7.1	47.8	1.7	6.6
Mexico	0.5	1.2	1.2	8.9	8.9
Morocco	30.0	23.3	21.8	48.4	16.0
Peru	0.6	0.1	0.2	14.5	14.5
Portugal	50.0	43.9	38.5	20.4	17.7
Spain	188.7	446.7	348.0	168.0	77.4
Syria	26.5	49.7	47.7	32.6	33.9
Tunisia	56.0	96.7	45.1	8.5	7.9
Turkey	82.0	111.7	88.5	123.5	117.8
Unit. States	2.2	0.8	27.0	68.1	108.0
Yugoslavia	4.5	3.0	4.0	0.5	1.5
Rest of world	10.7	4.1	58.4	11.0	96.9
TOTAL World	805.0	1,640.5	1,584.0	726.5	708.5

TABLE 1.3

Production of olive oil in EEC countries for the "olive year" 1987/88 (Thousand metric tons).

Country	Production
Spain	770.0
Italy	742.5
Greece	321.7
Portugal	38.0
France	4.0
TOTAL	1,876.2

TABLE 1.4

World olive oil production for the year 1981-90 (Thousand metric tons)*.

Years	Production
1981/82	1,583
1982/83	2,160
1983/84	1,444
1984/85	1,580
1985/86	1,480
1986/87	1,540
1987/88	1,710
1988/89	1,550
1989/90	1,850

*U.S. Department of Agriculture. Foreign agricultural service oilseeds and products (11).

A real effort for the development of olive cultivation in South Africa started in 1925 although the first olive plants were brought to the Cape in 1661. Recently, there has been significant interest for developing olive production in China, India, and other countries. Figure 1.1 shows where olive trees are cultivated today, all over the world.

World Production and Consumption

There are about 805 million olive trees in the world, covering approximately 24 million acres. Almost 98% of these trees are growing in the

Mediterranean area. The annual world production of olive fruit is about 9.4 million metric tons (MT). Approximately 720,000 MT of this production are consumed as table olives, and the remaining 8,680,000 MT are used for the extraction of olive oil. Nearly 1.6 million MT of olive oil are produced globally, plus 160,000 MT of olive pomace oil. About 260,000 MT of olive oil and 200,000 MT of table olives are traded in international markets.

The aggregate value of olive products is estimated to be 2.8 billion U.S. dollars- 2.5 billion represent the value of olive oil, and 0.3 billion the value of table olives. Olive products account for almost 25% of the farming income in the Mediterranean basin as a whole. The geographical distribution of olive oil production is shown in Table 1.1.

The production and consumption of olive oil and table olives in most of the olive-cultivating countries are shown in Table 1.2.

Table 1.3 gives the total olive oil production in European Economic Community Countries (EEC). Table 1.4 summarizes the total world olive oil production. The total world production of olive oil in comparison to other oils is shown in Table 1.5.

California produces 99% of the olives grown in the United States. Arizona is the only other state with a commercial production of olives. Most of the American olive crop is used for canning. The small amount of olive oil produced in California supplies less than 3% of the olive oil

TABLE 1.5

World production of edible oils (Thousand metric tons).*

Oils	1982/83	1983/84	1984/85	1985/86
Olive	2,160	1,444	1,580	1,480
Soya bean	13,616	13,160	13,330	13,640
Palm	6,006	5,589	7,040	8,290
Sunflower	5,809	5,699	6,080	6,380
Rapeseed	5,365	5,104	5,630	6,250
Cottonseed	3,346	3,364	3,870	3,430
Peanut	3,173	3,324	3,100	3,150
Cocoa butter	2,669	2,483	2,690	3,330
Palm kernel	731	769	950	1,130
Linseed	689	697	690	660
Fish	1,300	1,196	1,280	1,330

*U.S. Department of Agriculture. Foreign agricultural service oilseeds and products (11).

consumed in the United States, the rest is imported mostly from Spain and Italy. Major importing olive oil countries are: Australia, Brazil, Canada, France, Japan, Libya, Saudi Arabia, Switzerland, Turkey, United Kingdom, United States and Russia. Most of the world supply of olive oil originates from the Mediterranean countries. Of the 1600 thousand metric tons (MT) of olive oil produced globally in 1977, about 600 MT were produced in Italy, 360 MT in Spain, and 260 MT in Greece. Smaller quantities came from Tunisia, Turkey, Portugal, Morocco, Libya, and other countries near the Mediterranean Sea as well as from Argentina and California.

Olive oil is an important commodity in the daily diet of the Mediterranean people. The Greeks have the highest per capita consumption, 20.8 kg yearly, although it varies considerably among locations within Greece. The Spanish consumption is 10.0 kg/person/year followed by the Italian consumption at 8.1 kg/person/year (12,13). The consumption of olive oil in the United States has been increased considerably the last years.

References

1. Standish, R., The first of trees. The story of the olive. Phoenix House Ltd, London (1960).
2. Hurley, J., The tree, the olive, the oil in the old and new world. Trustee Albany College of Pharmacy, Trustee of Ray Brook. Tubercular Sanitarium Member State of Pharmacy (1919).
3. Hartmann, H.T. and P.C. Bougas, Olive production in Greece. Econ. Bot. 24:443 (1970).
4. Loukas, M. and C.B. Krimbas, History of olive cultivars based on their genetic distances. J. Hort. Science 58:121 (1983).
5. Goor, A., The place of the olive in the Holy land and its history through the ages. Econ. Bot. 20:223 (1966).
6. Camps-Fabrer, H., The olive tree and its economic significance in ancient north Africa. Olivae 2:9 (1984).
7. Kinman, C.F., Olive growing in the southwestern United States. United States Department of Agriculture Farmer's. Bulletin No 1249 (1922).
8. Hartmann, H.T. and K.W. Opitz, Olive production in California. Division of Agricultural Sciences, University of California. Leaflet 2474 (1977).
9. Hartmann, H.T., Olive growing in Australia. Econ. Bot. 16:31 (1962).
10. Fedeli, E., Lipids of olives. Prog. Chem. Fats and other Lipids 15:57 (1977).
11. U.S. Department of Agriculture. 1986, 1990. Foreign agricultural oilseeds and products (FOP G-86, G-90).

12. Katsoulis, N., M. Vagianou, and D. Karampakis, Olive oil today and after Greece will become member of EEC. Ministry of Agricul. Direction of Horticulture, Athens, Greece (In Greek) (1980).
13. Sellianakis, G., The effect on Greek olive cultivation due to the fact that Spain and Portugal will be members of EEC. Special edition of "Eleourgiki". Athens, Greece (In Greek) (1984).

Chapter 2

Botanical Characteristics of the Olive Tree

The botanical name of the olive tree is *Olea europea.* It originates from the Greek word for olive tree "elea" which became "olea" in Latin. The genus *Olea* is a member of the family *Oleaceae* which also contains several other well-known genera (1).

The olive tree is capable of living more than one-thousand years (2). Its longevity is related to its ability to send out shoots and roots from temporary buds generated by neoplastic formations which abound at the base of the lower part of the trunk. These formations usually are half-buried in the ground but highly developed and clearly visible in old trees.

The olive tree has an extensive root-system which tends to spread horizontally rather than vertically. Perhaps this is why it can even grow in rocky soils. When irrigated and in fertile soil this tree thrives vigorously. The wood of olive trees is resistant to decay and, if the top of the tree dies, a new trunk will often rise from the roots.

Many cultivars have been developed over the centuries differing in various ways including the size and color of the fruit and the oil content. Olive cultivars containing a low percentage of oil, and having high sugar content are usually used for table olives, while those with high oil content are used for oil production.

The olive tree has rather strict climatic requirements for successful tree growth and fruit production. It cannot be grown where winter temperatures drop below 15°F (-9°C) since such temperatures will kill the tree (3). Olive trees, however, must be subjected to a certain amount of chilling during the winter in order to initiate flowers for the following season's crop (3,4,5).

Different cultivars require different amounts of chilling. Certain table olive cultivars grown in Spain, Italy, and California require substantial periods of prolonged chilling to produce adequate flowering (4). Other cultivars grown in Northern Chile, in Southern Peru and in Greece (especially in Crete) will flower with considerable less chilling.

The olive tree is a broad evergreen tree. It varies in height, from 3 to 4 meters to as much as 20 meters, according to subspecies, race and environment and the foliage varies in proportion. The thick leathery leaves adhere to the tree for 2 to 3 years dropping mostly in the spring. Although as a rule the trees have a single trunk, there may be as many as four trunks. The bark, somewhere between greenish-gray in young trees, becomes dark gray as the trees grow older.

The flowers of olive trees are borne on a panicle inflorescence arising in the axils of the oppositely arranged leaves. They are usually produced laterally on branches formed the previous season, but may arise from dormant buds one or two years old. The small yellowish-white flowers

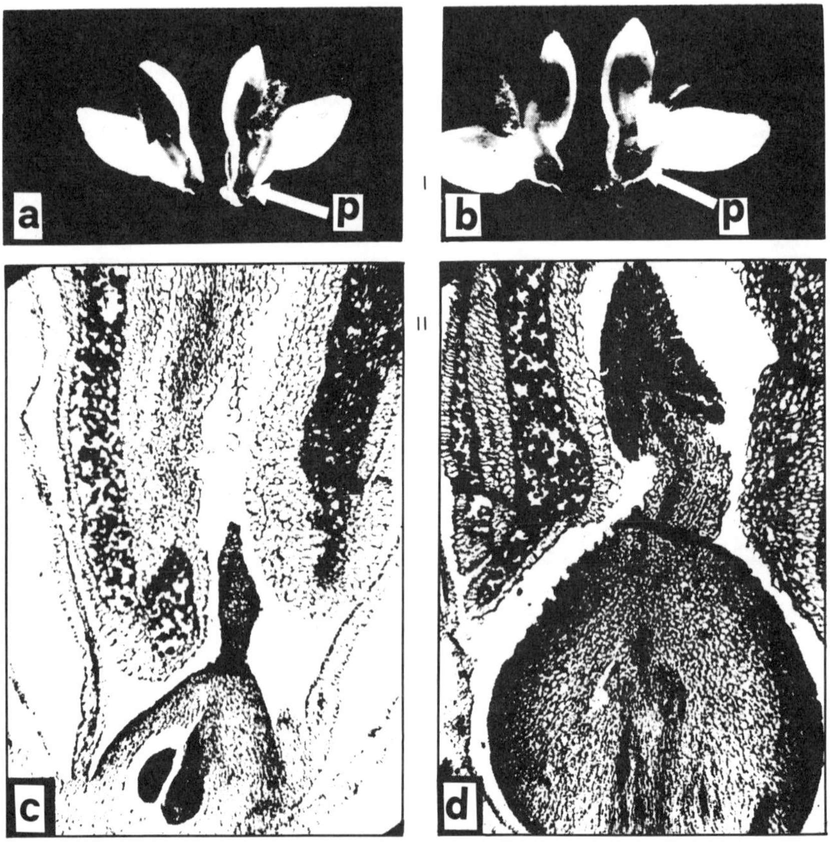

Fig. 2.1. I. Enlarged views of olive flowers at full bloom. (a) Staminate (male) flower (b) perfect (male and female) flower, (p) pistil. II. Photomicrographs of olive pistils (female part that develops into fruit) shortly before blooming (c) aborted pistil in staminate flowers (d) developed pistil in perfect flowers (1).

are of two types: (a) perfect, containing both stamens and pistil; and (b) staminate, containing only stamens, the pistils being abortive and non-functioning (1). The relative proportion of perfect and staminate flowers varies with cultivar and with the particular year. No purely pistillate flowers occur. The pistil is large in the perfect flower, almost filling the entire space within the floral tube; it is light green in the bud, and deep green at full bloom (Fig 2.1).

Alternate Bearing

Olive production often follows a two-year cycle with a good crop followed by a medium or poor crop. The alternate year fruit-bearing pat tern and the low yields are probably the biggest problems, the olive industry is facing. The alternate year fruiting is most frequent in cultivars with large-size fruits (e.g. Throumpolia, Tsounati).

Increase in olive production can be accomplished by causing an increase in the carbohydrate and nitrogen level (5). This can be done by irrigation, fertilization (adequate nitrogen), wide spacing of the trees and control of diseases (1).

References

1. Hartmann, H.T., and K. W. Opitz, Olive production in California. Division of Agricultural Sciences, University of California. Leaflet 1474 (1977).
2. Hurley, J., The trees, the olive, the oil in the old and new world. Trustee Albany College of Pharmacy, Trustee of Ray Brook, Tubercular Sanitarium Member State of Pharmacy (1919).
3. Hartmann, H.T., and P.C. Bougas, Olive production in Greece. Econ. Bot. 24:443 (1970).
4. Hartmann, H.T., and J. Porlingis, Effect of different amounts of winter chilling on fruitfulness of several olive varieties. Bot. Gazette. 119:102 (1957).
5. Sfakiotakis, E., Lectures on olive production. Aristotel Univer. of Thessaloniki, Greece (in Greek) (1987).

Chapter 3

Olive Fruit—Olive Oil Formation

The olive fruit, or olive, is a drupe (like the cherry or the peach), oval in shape, and consists of two main parts, the pericarp and the endocarp. The pericarp is composed of the epicarp, or skin; and the mesocarp, or pulp. The endocarp, (also called the pit or kernel) contains the seed (Fig. 3.1). The pericarp represents 66-85% of the weight of the fruit. The remaining is the endocarp. The seed does not exceed 3% of the weight of the fruit (1). The pericarp contains 96-98% of the total amount of oil; the remaining 2-4% of the oil is in the kernel.

Fresh, ripe olives are composed of water, oil, sugars, proteins, pectins, organic acids, tannins, oleuropein, inorganic components and so on. Citric, oxalic, malonic, fumaric, tartaric, lactic, acetic, and tricarballelic acids have been reported in the olive pulp. The olive fruit composition varies according to the type of cultivar, the environment and the degree of ripeness. The average chemical composition (%) of the fruit is shown in Table 3.1.

Cultivars with large fruits are known to have lower oil content and higher sugar content, and therefore are used as table olives. The olive cultivars best suited for oil production have medium-size fruits which contain 15-40% oil when ripe. Raina *et al.* (3) reported a maximum oil

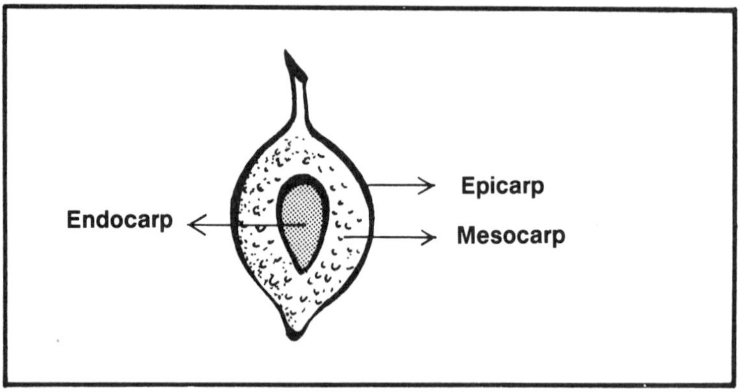

Fig. 3.1. Parts of the olive fruit.

TABLE 3.1

Chemical composition of olive fruit.

Component	Percent
Water	50.0
Oil	22.0
Sugars	19.1
Cellulose	5.8
Protein	1.6
Ash	1.5
TOTAL	100

content of 43.4% for the cultivar Coratina cultivated in India. Olives at the green stage used for pickling contain only a small amount of oil, as little as 8% or even less.

Olive oil cultivars have a pulp/kernel ratio of 4:1 to 8:1; table olive cultivars have a ratio of 7:1 to 10:1 (4). Some of the better known olive cultivars used for oil production are: Arauco, Corfolia, Koroneiki, Daphnolia, Fratoio, Lechin, Picual, Razzola, Rougette, Smertolia, Taggiasca, Tsounati, and Zorzalena. Cultivars preferred for table olive production include the following: Ascolano, Conservolia, Halkidikis, Gordal, Kalamata, Manzanilla, Mission, and Sevillano. Certain cultivars are suitable for both oil and table uses (e.g. Megaritiki, and Hohiblanco). Balatsouras (5) has described the most important olive cultivars.

Fruit Development-Maturation

Olive fruit exhibits a cyclic growth pattern. Growth is rapid during the first stage and slower during the second stage, in August and September. The third stage, in autumn, is again one of rapid growth and coincides with the color changes from green to yellow to red to black (6). Due to the fact that the value of olive fruits increases markedly with size, it is to the grower's advantage to delay, if possible, the harvest of edible olives (6). But it is unwise to delay harvest since the processing qualities of the fruit are affected by early frosts.

The increase of fruit size in autumn is related mainly to the increase in the moisture content of the fruit. If the trees lack soil moisture during this period, or if strong desiccating winds blow, the expected increase in fruit size cannot take place. With high transpiration and low soil moisture the leaves withdraw water from the fruits causing them to shrivel (6).

The rate of fruit development and maturation is affected by the following factors: (a) Age of olive tree-The fruit of younger trees mature faster since metabolism in young trees is much faster. (b) Cultivar-There are cultivars known for early maturity and others for late maturity. (c) The health stage of olive tree-Complete and normal maturity is a characteristic of healthy trees. On the contrary, sickly trees have difficulty bringing their fruit to maturity resulting in fruits of small size, poor appearance and with low olive oil content. and (d) Ecological factors-Moisture of the soil is very important since dry spells retard fruit maturation. Moisture is necessary during the fruit development. If there is enough moisture during that period the maturation process goes well and the oil is rich in tripalmitin and triolein (7). Light significantly affects olive fruit maturation. In areas with a lot of sunlight, maturation is rapid. Light affects not only the process of fruit maturation but also the quality of oil. Ample sunshine is related to the presence of many aromatic constituents in the oil (7). The type and the quantity of fertilizers applied to the soil affect fruit maturation.

The weight of a single fruit may vary at maturity from 2 to 12g, although the extreme values of 0.5 or 20g may be reached occasionally. During the growth period of the olive fruit (June to November for the

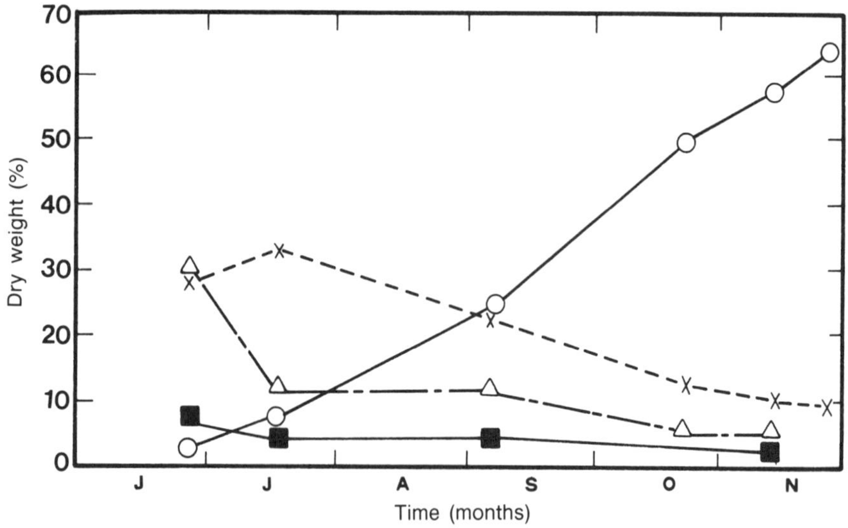

Fig. 3.2. Changes in certain olive flesh constituents during maturation of cultivar Gordal (O:oil,X: reducing sugars, Δ: fiber, ■: protein (Nx6.25) (8).

northern hemisphere), along with the increase in weight, several other changes occur (Figure 3.2). In the cultivar Gordal, as an example, and on dry weight basis, the oil content of the fruit increased continually; the reducing sugar content, after a brief rise, declined up to maturity; the crude fiber content dropped sharply initially and gradually afterward; the protein content (Nx6.25) remained low throughout the development of the fruit, as did the ash content (not shown in Fig. 3.2) (8).

Constituents of the Olive Fruit

The main constituents of the olive fruit are water, sugars, proteins, anthocyanins, and oleuropein.

Water

Water is one of the main constituents of the olive fruit and accounts for about 70% of its weight. Sufficient water in the soil results in large fruit size, and prevents fruit shriveling. Water serves as a solvent for the organic acids, the tannins, oleuropein, and the other water-soluble com-

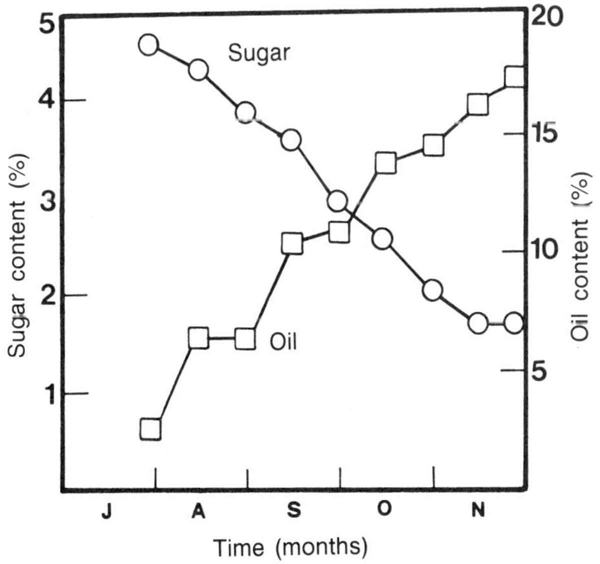

Fig. 3.3. Changes in oil and total sugar contents of the olive flesh during maturation.

ponents of the fruit. The amount of water found in the fruit depends on the stage of maturity, the cultivar and other factors such as climate, irrigation, and soil moisture. High oil content in the fruits implies low water content.

Sugars

The main sugars of the pulp are glucose and fructose. Sucrose, mannose, and galactose have also been found in certain cultivars (2). Glucose and, in smaller quantities, fructose were also found in the kernel of the fruit (9). Small quantities of sucrose present in the fruit diminish with matu-

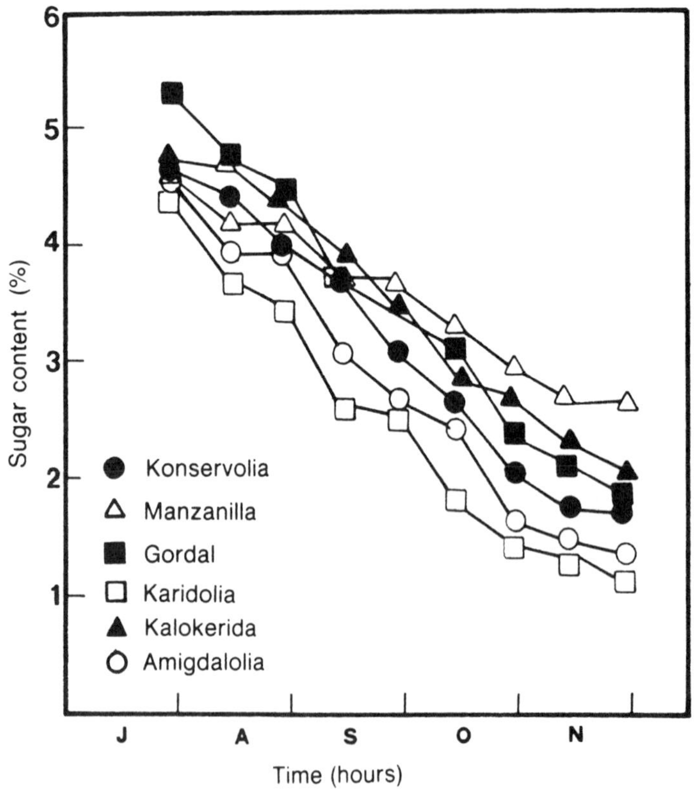

Fig. 3.4. Change in total sugar content of the flesh in certain cultivars during maturation.

rity. A decrease in the sugar content of olive fruit is related to an increase in the oil content (Fig. 3.3).

The amount of sugars present in olive fruit (used for Spanish-style table olives) is of special interest since during fermentation lactic acid is formed from these sugars. Lactic acid acts as a preservative and is responsible for the special taste of Spanish style table olives. The sugar content is diminished with the maturation. Figure 3.4 shows the change in the sugar content of six edible cultivars.

Proteins

Olive fruit mesocarp contains 1.5 to 3.0% protein depending on the cultivar and the maturity stage. All the amino acids present in other plant proteins were found in the olive fruit.

Protein content of the olive pomace was found to be 2-5% (9). Free nitrogen content varies considerably and constitutes less than 1% of the total nitrogen. Arginine, aspartic and glutamic acids constitute approximately 40% of the free amino acids found in the fruit of the cultivars Koroneiki, Throumbolia and Megaritiki. Arginine constituted approximately 25% of all essential amino acids followed by leucine and valine (10). Table 3.2 shows the amino acid composition of three olive fruit cultivars.

Anthocyanins

Anthocyanins and specifically the glycosides of cyanidin and peonidin are responsible for the purple and blue color of ripe olives (11-13).

The major pigments in eight cultivars of ripe olive fruit were found to be cyanidin-3-glucoside and cyanidin-3-rutinoside. Three other pigments were cyanidin triosides, probably derivatives of cyanidin-3-rutinoside, and cyanidin-3-sophoroside.

Luh and Mahecha (14) studied the anthocyanins of Manzanilla olives. The anthocyanins were identified as peonidin-3-monoglycoside acylated with p-coumaric acid, cyanidin 3-monoglycoside, and cyanidin-3-rhamnoglycoside acylated with caffeic acid.

Anthocyanins increase rapidly during the development of olives to reach a maximum, and then they decrease when the fruit is over-ripe. Light has a great influence in the pigment formation. Olive fruits normally developed contain ten times more anthocyanins than those ripened in the darkness (12).

TABLE 3.2

Comparison of amino acid composition in three Greek olive cultivars*.

	g/16 g nitrogen		
Amino acid	Koroneiki	Throumbolia	Megaritiki
Arginine	10.17	9.38	9.71
Histidine	2.74	2.99	2.34
Lysine	5.13	1.85	7.02
Methionine	1.29	1.52	1.16
Cystine	Trace	Trace	Trace
Phenylalanine	3.00	2.76	3.14
Tyrosine	2.42	2.52	2.22
Leucine	5.25	4.71	5.55
Isoleucine	3.04	2.96	3.28
Threonine	2.60	2.51	3.30
Valine	4.07	3.79	4.03
Alanine	5.45	6.49	3.60
Aspartic acid	10.99	11.11	12.28
Glutamic acid	10.72	12.54	11.85
Glycine	6.54	8.79	3.50
Proline	4.31	7.44	2.70
Serine	3.05	3.48	4.30

*Manoukas et al. (10)

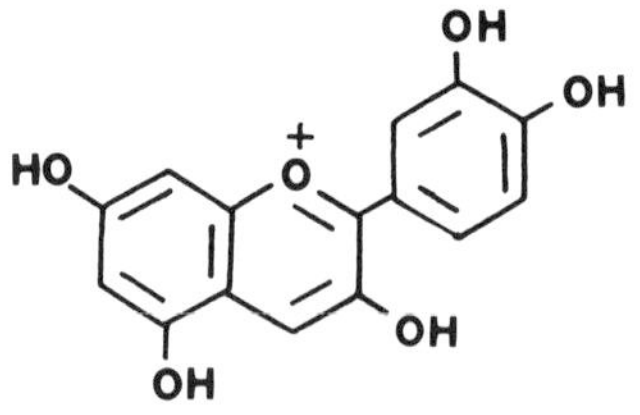

Cyanidin

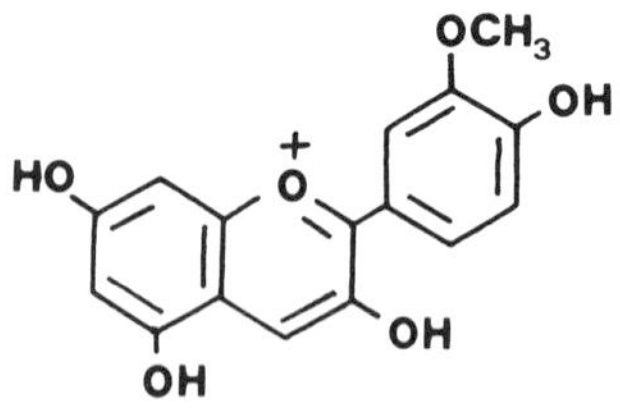

Peonidin

In the beginning the color of olives is green; it turns purple to bluish later, and becomes black in overripe or black pickled olives (Greek-style or California-style). The green color is due to chlorophylls, the purple and blue color to anthocyanins and the black color is formed by the oxidation of phenolic compounds including oleuropein (15).

Oleuropein

A phenolic glycoside typical of olives is oleuropein (16). It is responsible for the bitter taste of immature olives. As the fruit reaches maturity, the amount of oleuropein diminishes (9,11). This is why ripe olives are not as bitter a unripe ones. Amiot *et al.* (17) reported that the oleuropein content of olive fruits could reach 14% of the dry matter in young fruits and it should be seriously considered in the processing of green olives. Small fruit cultivars are characterized by high oleuropein content whereas large fruit cultivars usually contain small amounts of oleuropein. Oleuropein is water-soluble, and remains in the aqueous phase during the olive oil extraction process. During the processing of Spanish-style edible olives, oleuropein is hydrolyzed by a sodium hydroxide solution (1-2%) at room temperature. Removal of the excess alkali with successive rinsings does not cause a return of the bitterness.

Oleuropein is also hydrolyzed by acids, and by the enzyme glycosidase. Oleuropein gives many reactions which coincide with those of o-diphenols. The Hoepfner test is based on a sensitive reaction which shows the presence of o-diphenols, and helps in detecting olive oil in other oils (9).

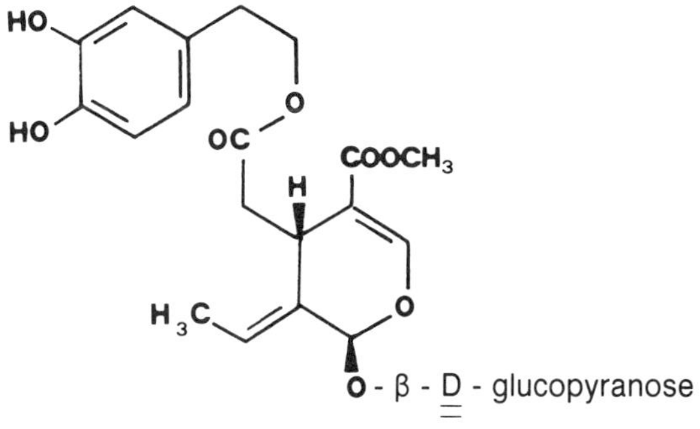

Fig. 3.5. The structure of oleuropein.

Oxalic Acid

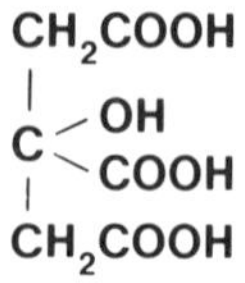
Citric acid

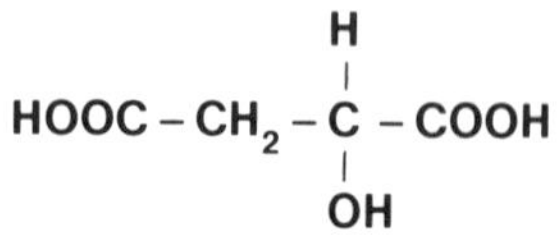
Malic acid

TABLE 3.3

Inorganic nutrients determined in the olive fruit mesocarp of three stages of maturity (mg/100g fresh fruit)*.

	Olive fruit mesocarp		
Element	Stage I	Stage II	Stage III
Na	4.97	4.57	3.20
K	322.69	406.07	457.19
Ca	40.92	36.32	33.15
Mg	18.86	19.82	12.49
Mn	0.19	0.18	0.13
Fe	1.43	1.45	1.22
Zn	0.78	0.62	0.71
Cu	0.01	0.03	0.01
P	39.74	42.60	51.13

*Manoukas et al. (18)

TABLE 3.4

Proximate analysis of the olive fruit mesocarp in three cultivars (g/100g fresh weight)*.

	Cultivars		
Constituent	Koroneiki	Throumbolia	Megaritiki
Moisture	72.5	77.2	70.7
Lipids	16.8	4.3	16.1
Ash	1.2	1.0	1.3
Nitrogen (total)	0.32	0.19	0.20
Nitrogen (free)	0.003	0.0004	0.001

*Manoukas et al. (10)

TABLE 3.5

An analysis of the olive seed (g/100g fresh weight)*.

Constituent	Content	Constituent	Content
Lipid	37.40	Phosphorus	0.32
Carbohydrate	33.30	Calcium	0.20
Protein (% N x 0.25)	21.94	Iron	0.02
Moisture	4.60	Magnesium	0.02
Ash	2.20		

*Samir and Hegsted (19)

Oleuropein is also known for its pharmaceutical properties (2). The structure of oleuropein is shown in Figure 3.5.

Other constituents of the olive fruit

The olive pulp also contains steroids, cerebrosides and sulfolipids. Small quantities of the following organic acids have been reported to be present in olive pulp: citric, malic, oxalic, malonic, fumaric, tartaric, lactic, acetic and tricarballelic (2). The olive pulp content in certain acids is as follows: Oxalic acid, 0.10-0.17%; citric acid, 0.1-0.15%; and malic acid, 0.01-0.70%. These acids are either free or in the form of salts (9).

Certain minerals such as iron, calcium and potassium are also found in the pulp of olive fruits. The results of mineral analysis of olive fruit mesocarp at three stages of maturity are shown in Table 3.3. It is worth noting that the potassium content is higher than the total content of all other minerals put together. Table 3.4 gives the composition of olive fruit mesocarp for three Greek cultivars.

The result of a proximate analysis along with the content in certain minerals of the olive seed appears in Table 3.5.

Olive Oil Formation

The biosynthesis of the olive oil triglycerides involves the following three processes (20).

1. Fatty acids are synthesized by successive additions of malonyl CoA to a primer molecule of acetyl-CoA. Since the three carbon malonyl residue is decarboxylated after each addition, the progressive lengthening of the fatty acid chain by two carbon atoms is understandable. A multi enzyme system catalyzes the condensation, reduction, and

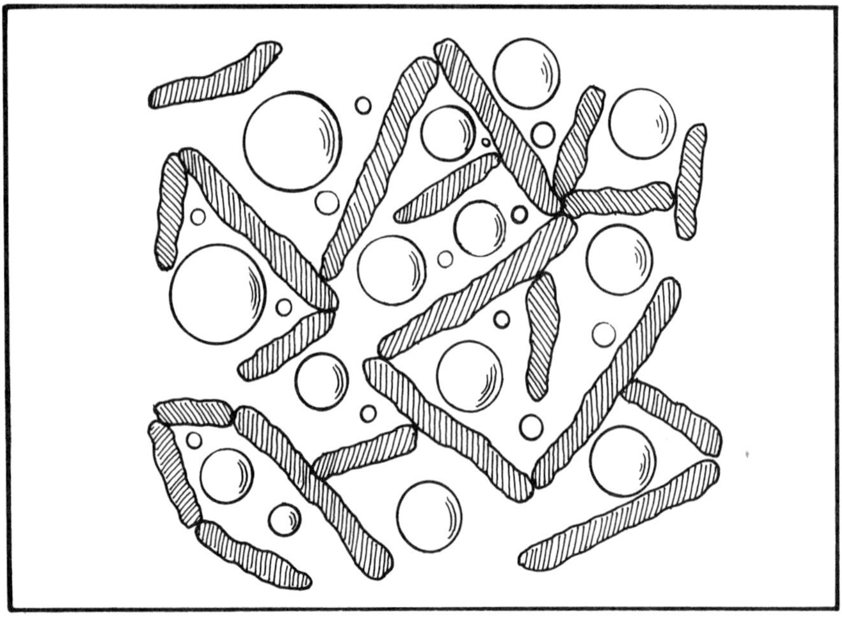

Fig. 3.6. Olive oil drops in the cells (21).

dehydration reaction that are necessary for the completion of the fatty acid synthesis.

2. Glycerol phosphate is formed from the dihydroxy acetone phosphate of the glycolytic pathway.
3. Fatty acids, as CoA derivatives, are then transferred to the free hydroxy groups of the glycerol phosphate. Dephosphorylation and completion of the esterification of glycerol follow.

Olive oil is formed during the following four stages:

1. Initial stage: Along with the fruit growth a small amount of oil is formed.
2. Stage of major concentration: During this period almost all of the olive oil is synthesized.
3. Stable stage: The olive oil content in the pulp does not change.
4. Decrease stage: A decrease in olive oil content is the result of over maturity.

The formation of olive oil in the pulp starts by the middle of July. The amount of oil increases gradually through fall and reaches its maximum in late December or January (6). Initially, olive oil is dispersed within the fruit cells forming drops which grow in size, and eventually may fill the entire cell. Olive oil drops vary from 39 to 63 μ in diameter. Bigger drops can deform the cell by pressing the cell membrane. Many times a big drop is visible along with smaller ones in the same cell (Fig. 3.6).

References

1. Frezotti, G. and M. Manni, Olive oil processing in rural mills. FAO, Rome. Agricultural Development paper No 58 (1956).
2. Fedeli, E., Lipids of olives. Prog. Chem. Fats and other Lipids. 15:57 (1977).
3. Raina, B.C., A.K. Bhatia, S.K. Katiyar, and A.K. Gupta, Quality characteristics of promising Himachal olive varieties. (Olea europaea C.). J. Food Sci. Techn. 23:237 (1986).
4. Sfakiotakis, E., Lectures on olive production. Aristotel Univer. of Thessaloniki, Greece (In Greek) (1987).
5. Balatsouras, G., The olive tree. Mauromatis, N. Editor. Athens, Greece (In Greek) (1984).
6. Hartmann, H.T. and K.W. Opitz, Olive production in California. Division of Agricultural Sciences, University of California. Leaflet 2474 (1977).
7. Papanastassiou, D. Modern olive oil processing. The Technology of olives. Spiros, Spirou, and Son, Editor. Athens, Greece (In Greek) (1966).
8. Fernandez-Diez, M.J., The Olive. In Biochemistry of fruits and their products. Hulme, A.C. Editor. Academic Press, Inc., N.Y. (1971).
9. Aligizakis, M., Processing and canning of edible olive. Mauromatis, N. Editor. Athens, Greece (In Greek) (1982).
10. Manoukas, G., B. Mazomenos and M. Patrinou, Amino acid compositions of three varieties. J. Agr. Feed Chem. 21:215 (1973).
11. Kiritsakis, A. and P. Markakis, Olive oil, A review. Advances in Food Research 31:453 (1987).
12. Luh, B.S. and C.E. Kean, Canning of fruits. In Commercial fruit processing. Woodroof, J.G. and B.S. Luh, Editors. The AVI Publishing Company, Inc., Westport Connecticut (1975).
13. Timberlake, C.F. and P. Bridle, Distribution of anthocyanins in food plants. In Anthocyanins as food colors. Markakis, P. Editor. Academic Press, Inc. N.Y. (1982).
14. Luh, B.S. and G. Mahecha. Anthocyanins in manzanilla olives. J. Chinese Agr. Chem, Soc., Sp. Iss, 1-24 (1971).
15. Vaughn, R.M., K.C. Simpon, and C.O. Chicester, Compounds responsible for the color of black ripe olives. J. Food Science. 26:227 (1961).
16. Cruess, W.Y., Commercial fruit and vegetable products. 4th edition. McGraw-Hill Book Co., N.Y. (1958).
17. Amiot, M., A. Fleuriet and J. Macheix. Importance and evolution of phenolic compounds in olive during growth and maturation. J. Agr. Food Chem.

34:823 (1986).

18. Manoukas, G., A. Grimanis, and B. Mazomenos, Inorganic nutrients in natural and artificial food of Dacus oleae larva (Diptera: Tephritidae). Ann. Zool. Ecol. Anim. 10:123 (1978).
19. Samir, M. and D.M. Hegsted, Olive kernels as a source of protein. Nutr. Rep. Int. 9:117 (1974).
20. Hess, D. Plant Physiology. Springer-Verlag. N.Y. (1975).
21. Moreno Martinez, J.M., Estudios fisicoquimicos sobre las pastas de aceitunas molidas. Heterogen, en la distribucion de los solidos de la pasta. Grasas y Aceites. 16:17 (1965).

Chapter 4

Chemistry of Olive Oil

Olive oil is mainly composed of triglycerides and contains small quantities of free fatty acids (FFA), glycerol, phosphatides, pigments, carbohydrates, proteins, flavor compounds, sterols and unidentified resinous substances (1-4).

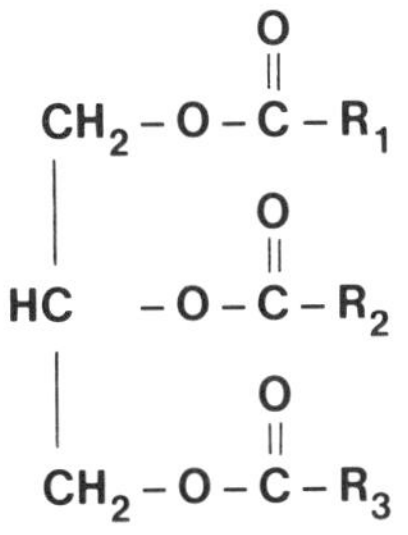

Triglyceride

Olive oil constituents can be divided into two categories: the saponifiable fraction (triglycerides, FFA, phosphatides, etc.), and the unsaponifiable fraction (hydrocarbons, fatty alcohols, etc.). In virgin olive oil the unsaponifiable constituents account for 0.5-1.5% of the oil while in olive pomace oil (see Chapter 8) the unsaponifiable fraction is about 2.5%.

The Fatty Acids of Olive Oil

The fatty acid composition of olive oil varies widely depending on the cultivation, the climate conditions of the area and several other factors (5-9). Marzouk *et al.* (10) studied the effect on the fatty acid composition, during ripening of the olive fruit, of salt in irrigation water. They reported that increasing NaCl concentration caused a progressive decrease in fatty acid synthesis in the olive fruits. Oleic acid increased during ripening but decreased percentages as the NaCl concentration was increased. No significant changes in the palmitic and linoleic acid levels were observed. Most of the fatty acids present in olive oil are unsaturated. The major fatty acids present as glycerides in olive oil are oleic (18:1), linoleic (18:2), palmitoleic (16:1), palmitic (16:0), and stearic (18:0) (1, 11-13). Oleic acid is present in much higher concentrations (70-80% in weight) than the other acids.

The fatty acid composition of olive oil is shown in Table 4.1. Gracian (11) reported the fatty acid composition of olive oil based on the analyses of about 4,000 samples from Spain, Italy, Greece, Tunisia, Argentina, and the United States. The Codex Alimentarius Committee on Fats and Oils (14) has set the following low and high limits for the three principal acids of olive oil: oleic, 50-83%, palmitic, 7-20%, and linoleic, 3-20%. The wide rages of fatty acids in Table 4.1 are probably related to both genetic factors and environmental conditions during the development of the fruit (1,7,9). Colakoglu (15) reported small amounts of a C_{24} fatty acid in olive oil. The presence of elaidic acid (C18:1,trans) in olive oil has not been positively determined. Kiritsakis et al. (16) detected traces of elaidic acid in olive oil by means of infrared absorption spectrophotometry. Traces of lauric, myristic, arachidic and eicosanoic acids have been reported in Indian olive oil (17). Rana and Ahmed (18) found the oleic acid content in Libyan olive oil to be 43.7% which is very low, while the linoleic acid content was over 30% which is unusually high.

Delay in harvesting tends to increase the content of unsaturated fatty acids, especially linoleic, at the expense of palmitic acid. Unsaturated fatty acids also increase when the temperature of the growing area is low.

TABLE 4.1

Distribution of fatty acids in olive oil (%)*.

Acid	Content
Oleic acid	56.0-83.0
Palmitic acid	7.5-20.0
Linoleic acid	3.5-20.0
Stearic acid	0.5- 5.0
Palmitoleic acid	0.3- 3.5
Linolenic acid	0.0- 1.5
Myristic acid	0.0- 0.1
Arachidonic acid	0.8 (max)
Behenic acid	0.2 (max)
Lignoceric acid	1.0 (max)
Heptadecanoic acid	0.5 (max)
Heptadecenoic acid	0.6 (max)

*International Olive Oil Council (12)

Iverson and Firestone (19) categorized two types of oils based on fatty acid composition: one has low linoleic, low palmitic and high oleic acid content; the second type of oil is relatively rich in linoleic and palmitic acids but contains little oleic acid.

Viola (20) considered the ratio of vitamin E to polyunsaturated fatty acids (PUFA) in olive oil to be more favorable to health than the ratios in other vegetable oils. Several researchers (9,21-23) consider that olive oil has important beneficial physiological effects (see Chapter 17).

Olive fruit from cool areas gives oil with more unsaturated fatty acids than the fruit produced in dry and warm areas. Olive oil from north Mediterranean countries is richer in liquid glycerides compared to the oil from south Mediterranean countries which is richer in solid glycerides (mainly glycerides of palmitic and stearic acids). "Thick" olive oil, as that produced in warm areas, is often winterized by chilling for removal of certain saturated glycerides (see Chapter 9).

The typical fatty acid composition of olive oil from various countries is presented in Table 4.2. Table 4.3 gives the fatty acid composition of olive oil and other vegetable oils. Olive oil contains more oleic acid and less linoleic and linolenic acids than other vegetable oils. This renders olive oil more resistant to oxidation (see Chapter 12).

Olive oil differs from the majority of other edible oils in the following aspects (25):

- Contains high percentage of monounsaturated fatty acids (oleic).

TABLE 4.2

Fatty acid composition of olive from different countries*.

	Fatty Acids (%)						
Country	Saturated				Unsaturated		
	C14:0	C16:0	C18:0	C20:0	C16:1	C18:1	C18:2
Greece	-	7.5-16.0	1.4-3.8	-	0.5-2.3	57.6-93.5	1.6-23.6
Italy	-	7.1-17.5	0.3-3.4	-	0.2-5.5	64.1-85.0	1.0-15.0
Spain	0.2	9.5	1.4	0.2	-	81.6	7.1
Argentina	-	9.8-20.0	0.3-2.9	-	0.2-3.4	54.0-79.1	5.3-22.7
Tunisia	-	13.9-21.1	1.3-2.5	-	1.0-2.2	55.2-70.6	9.5-20.1
Israel	0.5	10.0	3.3	0.1	-	77.5	8.6
USA	-	-	7.0	2.3	0.1	85.8	4.8

*Gracian (11)
Bailey's Industrial Oil and Fat Products (24)
Ministry of Commerce. Studies on Greek Olive Oil (8)

TABLE 4.3

Fatty acid composition of olive oil and other vegetable oils.*

Fatty acids	Olive oil	Corn oil	Safflower oil	Colza oil	Peanut oil	Palm oil	Sun-flower oil
14:0	traces	-	traces	-	traces	1	-
16:0	14	13	8	4	6	48	11
18:0	2	traces	3	2	5	4	6
20:0	traces	traces	traces	-	2	-	-
22:0	-	traces	-	-	3	-	-
24:0	-	-	-	-	1	-	-
14:1	-	-	-	-	-	-	-
16:1	2	-	-	-	traces	-	-
18:1	66	29	13	19	61	38	29
20:1	-	-	-	13	-	-	-
22:1	-	-	-	40	-	-	-
18:2	16	54	76	14	22	9	54
18:3	-	-	-	8	-	-	-

*Fedeli and Jacini (26)

- Contains low percentage of saturated fatty acids.
- Has high concentration of minor compounds, especially the virgin olive oil.

The Glycerides of Olive Oil

Most of the fatty acids of olive oil are present as triglycerides. Fedeli (1) reported the major triglycerides as: POO (18.4%), SOO (5.1%), POL (5.9%), OOO (43.5%), OOL (6.8%), (P=palmitic, O=oleic, S=stearic, and L=linoleic acids). Table 4.4 shows the percentage distribution of individual triglycerides of olive oil. The determination of palmitic acid in position-2 of olive oil glycerides can be used for detecting virgin and refined olive oil adulteration (see Chapter 15).

Nonglyceride Constituents of Olive Oil

Several minor nonglyceride compounds are present in olive oil. Table 4.5 gives the percentage content of these minor compounds present in olive oil and other vegetable oils. Table 4.6, on the other hand, summarizes the minor components determined in virgin and refined olive oil. A comparison of the unsaponifiable fraction of virgin olive oil and olive pomace oil is given in Table 4.7.

TABLE 4.4

Fatty acid distribution among the chief triglycerides of olive oil (%).*

Number of double bonds per Triglyceride	Fatty acid distribution	Content
0	0	0.0
1	POP	2.9
	PPO	0.6
	POS	0.5
	PSO	0.3
	SOS	0.2
	N.D.	0.2
2	PPL	1.2
	POO	18.4
	POS	2.3
	PLS	0.1
	LPS	0.7
	SOO	5.1
	N.D.	2.2
	PLO	0.2
3	POL	5.9
	OPL	0.9
	SLO	2.4
	LSO	0.7
	SOL	1.3
	OOO	43.5
	N.D.	0.6
	PLL	0.4
4	LPL	0.2
	OOL	6.8
	OLO	3.5
	SLL	0.2
	LSL	0.3
	N.D.	0.1

*Fedeli and Jacini (26) (P, palmitic; O, oleic; S, stearic; L, linoleic; N.D., nondetermined acid).

The nonglyceride fraction of olive oil contains nonglyceride fatty acid esters, hydrocarbons, sterols, triterpene alcohols, tocopherols, phenols, phospholipids, chlorophylls, and flavor compounds. Quantitative analyses of certain minor compounds such as sterol and triterpenic alcohols

TABLE 4.5

Unsaponifiable composition of some well known vegetable oils (%).*

Oils	Hydro-carbons	Squalene	Aliphatic alcohols	Terpenic alcohols	Sterols
Olive	2.8 - 3.5	32 - 50	0.5	20-26	20 - 30
Linseed	3.7-14.0	1.0 - 3.9	2.5-5.9	29-30	34.5 - 52
Tea-seed	3.4	2.6	-	-	22.7
Soybean	3.8	2.5	4.9	23.2	58.4
Rapeseed	8.7	4.3	7.2	9.2	63.6
Corn	1.4	2.2	5.0	6.7	81.3

*Jacini et al. (27)

TABLE 4.6

Minor components of virgin and refined olive oil (ppm).*

	Oils	
Components	Virgin olive oil	Refined olive oil
Hydrocarbons	2000	120
Squalene	1500	150
β-Carotene	300	120
Tocopherols	150	100
Phenols and related substances	350	80
Esters	100	30
Aldehydes and ketones	40	10
Fatty alcohols	200	100
Terpene alcohols	3500	2500
Sterol alcohols	2500	1500

*Fedeli (25)

are indicators of the authenticity of olive oil. The minor constituents in the unsaponifiable fraction of olive oil were studied. Some of these constituents may be destroyed during the saponification process.

A more complete recovery of the nonglyceride constituents in olive oil may be accomplished by the crystallization of glycerides from a polar solvent such as acetone or ethyl acetate at low temperatures (-15 to 60°C). The nonglycerides remaining in the mother liquor can be isolated and then determined by chromatographic and other analytical methods (1).

TABLE 4.7

Composition of the unsaponifiable fraction of olive oil and olive pomace oil (ppm).*

Constituents	virgin olive oil	Pomace olive oil
Squalene and other hydrocarbons	3000-5000	1500
Sterols	1500	2500
Triterpenoid alcohols	1000	1000
Higher fatty alcohols	-	1500
Carotenoid pigments, tocopherols, antioxidants, volatilc and other compounds	2500-4500	3500

*Gracian (11)

TABLE 4.8

Distribution of fatty acids among three groups of nonglyceride esters in olive oil (%).*

Fatty acid	Group A	Group B	Group C
14:0	-	2.2	0.2
16:0	10.6	5.9	5.5
16:1	3.0	1.8	0.6
18:0	0.5	0.7	0.4
18:1	49.5	41.4	89.4
18:2	22.9	26.8	2.5
18:3	0.7	0.5	0.3
20:0	1.4	9.0	0.5
20:1	8.5	9.5	0.2

*Fedeli (1). Esters with Group (A) long chain alcohols, sterols, and some triterpene alcohols; Group (B) most triterpene alcohols; or Group (C) methanol and ethanol.

Nonglyceride Esters of Fatty Acids

Although the majority of fatty acids are esterified with glycerol, small quantities of fatty acids form esters with other alcohols such as methanol, ethanol, sterols, and triterpene alcohols. Jacini *et al.* (27) reported the detection of esters of sterols such as β-sitosterol, campesterol, stigmasterol; and triterpenic alcohols such as cycloartenol, or 2,4, methylene cycloartenol. The distribution of fatty acids in the esters of different alcoholic groups is shown in Table 4.8.

Hydrocarbons

Squalene which is a biochemical precursor of sterols is the major hydrocarbon of olive oil. Hamann *et al.* (28) reported that squalene is the major constituent (up to 40% by weight) of the olive oil unsaponifiables. Olive oil contains the largest amount of squalene among vegetable oils (29). The squalene content in certain oils is given in Table 4.9. The structure of squalene is shown in Figure 4.1.

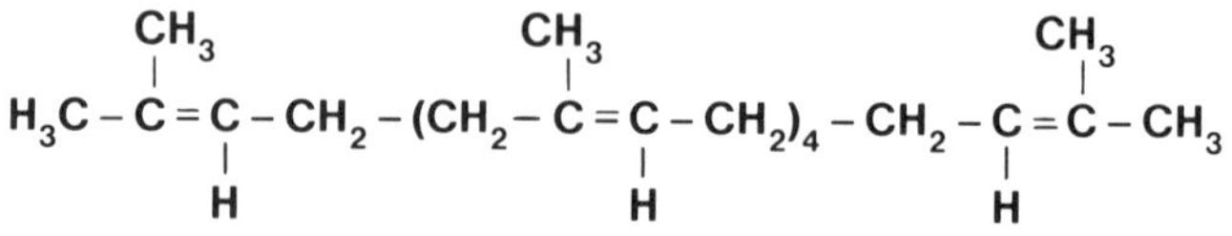

Fig. 4.1. The structure of squalene

TABLE 4.9

Squalene content of olive oil and other vegetable oils.*

Oil	Squalene (mg/100g)
Olive	136 - 708
Corn	19 - 36
Cottonseed	4 - 12
Peanut	13 - 49
Soybean	7 - 17
Sunflower	8 - 19
Teaseed	8 - 16
Sesame	3
Rape	28
Mustard	7
Rice bran	332
Grapeseed	7
Almond	21
Cocoa	none
Coconut	2
Linseed	4
Butter	7
Cod liver	31
Chicken	4
Lard	3
Beef	10

*Mehlenbacher (33)

The polycyclic aromatic hydrocarbons in olive oil are phenanthrene, pyrene, fluoranthrene, 1,2-benzanthracene, chrysene, and perilene (1). Both even and odd carbon-numbered n-paraffins, ranging from C_{11} to C_{30} and branched hydrocarbons were found in olive oil (1,30). Ciusa and Morgante (31) determined 14 polycyclic aromatic hydrocarbons in olives. The hydrocarbon content is higher in unripe than in ripe olives. It is believed that hydrocarbons are the result of natural metabolism.

Stancher *et al.* (32) determined both polar and nonpolar carotenoids in olive oil by GLC. Two major carotenoids, lutein and β-carotene were present in concentrations which varied according to the type of oil and the extraction process applied.

β-Carotene (Fig. 4.2) is present in olive oil in very small quantities (0.3-3.6 mg/kg). Zonta and Stancher (34) reported that the concentration of lutein and β-carotene, in olive oils vary according to the oil quality from 0.153-0.444 and 0.085-0.496 (mg/100g), respectively.

It has been shown that β-carotene quenches singlet oxygen (see Chapter 12) and thereby inhibits lipid oxidation by singlet oxygen (3,35-59).

Sterols

Fedeli (1) reported that Italian olive oils contain β-sitosterol accounting for 96% of the total sterols, campesterol representing 3% and stigmasterol accounting for the remaining 1% of the total sterols. Boskou and Morton (40) found anemasterol to represent 8% of total sterols content in Greek olive oil. Itoh *et al.* (41) reported the presence of the aforementioned four sterols in olive oil. Leone *et al.* (42) showed that the peroxide value of olive oil increased and the sterol content decreased during storage.

Olive oil contains four methylsterols which are (1) 4α-methyl-24-methylene Δ^7 cholestene-3 β-ol, (2) 4α-methyl-24-methyl-Δ^7 choles-

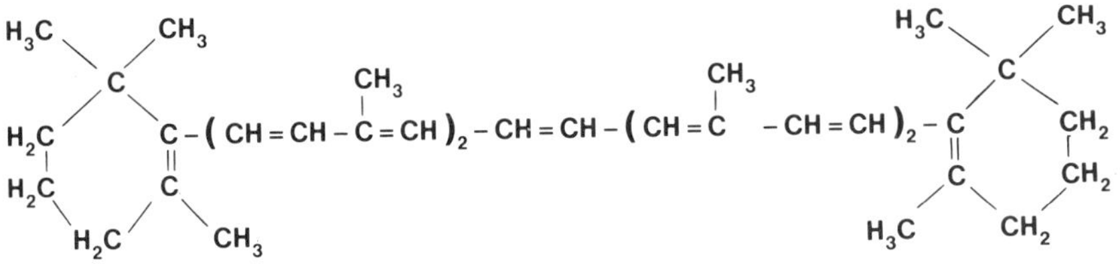

Fig. 4.2. The structure of β-carotene.

tene-3 β-ol, (3) 4α-methyl-24-ethyl-Δ^7 cholestene-3 β-ol and 4α-methyl-24-ethylidine- Δ^7 cholestene-3 β-ol. Kiosseoglou *et al.* (43) found that olive oil contains 4-methyl-and 4,4-dimethyl-sterols as esters with fatty acids. Boskou (44) reported that 4α-methyl-sterols are more susceptible to oxidation compared to the common sterols and triterpene alcohols.

Itoh *et al.* (41) determined four sterolic fractions: triterpene alcohols, 4-methylsterols, sterols and triterpene dialcohols in Spanish olive oils. Four components of these fractions were considered as new sterols: 24-methylene-31-nor-9(11)-lanosterol, 24-methyl-31-nor, -23 dehydroxy-cloartanol, 24-ethyl-23-dehydrolophenol and 5,-23-stigmastadienol.

Meneghetti *et al.* (45) analyzed the sterol fractions of olive oil as acetyl derivatives, separated on TLC by electronic ionization GLC-MS. Gutfinger and Letan (29) reported that the sterols in olive oil extracted from olive pulp and olive kernel differ quantitatively but not qualitatively. Table 4.10 shows the average composition of the sterol fraction of virgin, refined and olive pomace oil.

The total sterol content of olive oil is in the range of 180-265 mg/100g. Leone *et al.* (42) reported that the sterol content of olive oil is reduced due to oxidation during storage and processing. Tiscornia and Bertini (46) reported that the sterol content is not related to the acidity of the oil. Cottonseed oil, avocado oil and olive oil contain the same sterols; β-sitosterol represents 90% of total sterol content in these oils. Excessive amounts of stigmasterol indicate the presence of soybean oil in olive oil (29).

TABLE 4.10

Sterol composition of virgin, refined and olive pomace oil.*

Compounds	Composition (%)		
	Virgin Oil	Refined Oil	Olive Pomace Oil
Campesterol	3.0	3.3	3.7
24-Methylenecholesterol 7-22-ergostadienol	0.4	0.5	0.3
Stigmasterol	0.7	3.0	2.3
Sitosterol 5-23-stigmastadienol	89.3	87.2	91.0
Fucosterol	0.3	0.3	0.5
28-Isofucosterol	5.6	5.2	1.7
5.24(25)-Stigmadieno φ	0.1	0.4	0.5

*Itoh *et al.* (41)

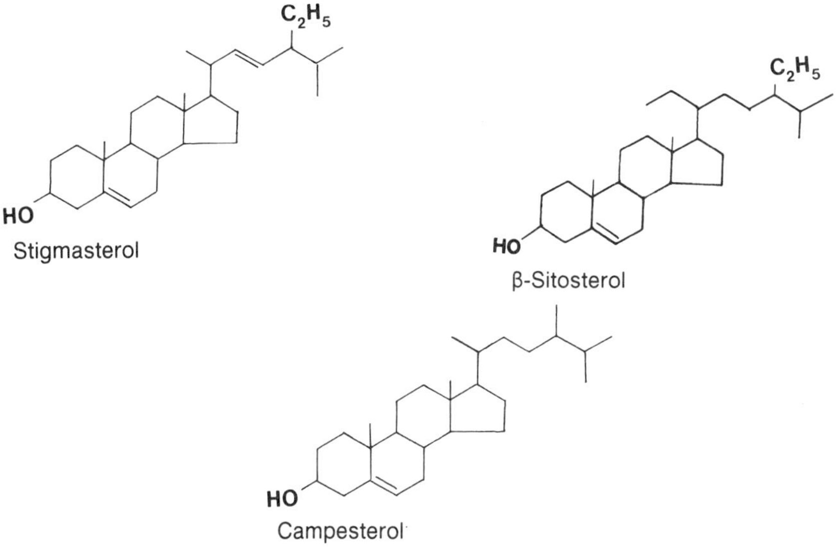

Fig. 4.3. The structures of the main sterols in olive oil.

Lotti *et al.* (47) studied the sterol composition of two Italian cultivars grown at seven locations, and found that the sterol composition was little affected by the climatic conditions. Paganuzzi (48) determined the sterol composition of four olive cultivars using GLC and reported that citrostadienol was the major component, followed by cycloeucalenol and/or gramisterol and obtusifol. He also found a component with relative gas chromatographic retention time corresponding to 31-norcycloartenol in pressure-extracted oil.

Morchio *et al.* (49) used capillary GLC to study the changes in total sterols in five classes of olive oil. They found that the total sterol content was directly related to the acidity of the oils (between 1 and 8%) and it decreased during the refining process. A wide range of sterol content from 850-1325 mg/kg in refined oil, to 3110-4650 mg/kg in hexane-extracted olive pomace oil was observed.

The structures of the main sterols found in olive oil are shown in Figure 4.3. Table 4.11 shows the sterol content of olive oil. The sterol content of olive pomace oil from different countries appears in Table 4.12.

TABLE 4.11

Sterol content of olive oil.

Sterols	Content (%)
Campesterol	2.0
Stigmasterol	0.5
β-Sitosterol	89.5
Δ 5-Anemasterol	8.0

TABLE 4.12

Sterol content of olive pomace oil from different countries.*

Country	Content (%)		
	β-Sitosterol	Stigmasterol	Campesterol
Italy	94.0	2.4	3.6
Spain	94.8	1.9	3.3
Tunisia	94.7	1.7	3.6

*Fedeli (1)

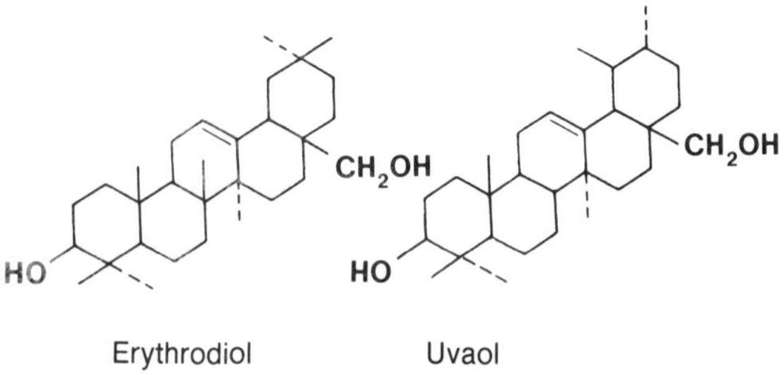

Fig. 4.4. The structures of erythrodiol and uvaol.

Triterpene Alcohols and Hydroxy Triterpenic Acids

Two dihydroxy triterpenes have been identified in olive oil: erythrodiol which is 3β-17β-dehydroxy- Δ^{12}-oleane (50,51) and uvaol which is 3β-11β-dihydroxy- Δ^{12}-ursene (1). Erythrodiol is present in olive oil at an average of 7.2% (50). The structures of erythrodiol and uvaol are shown in Figure 4.4.

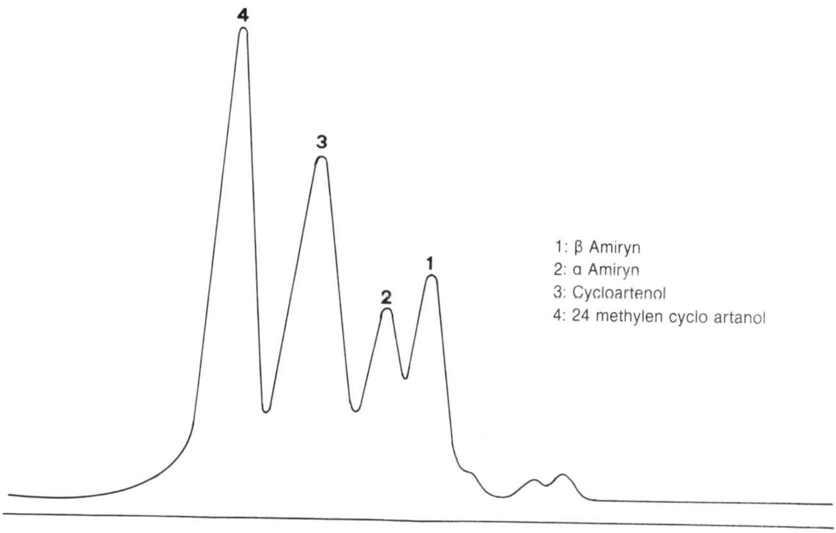

Fig. 4.5. Gas chromatogram of the triterpenic fraction of olive oil (1).

The presence of oleanolic acid (a triterpenic acid, of the α-amyrin series), as a consistent component of the olive fruit and of its oil, has long been known. More recent is the identification of maslinic, ursolic, and 2-dihydroxyursolic acids (1).

Paganuzzi (52) analyzed the triterpene alcohols of the unsaponifiable part of olive oils. He found α-and β-amyrin; euphol; taraxerol; dammaradienol; 24-methylene lanost,-8-enol; 24-methylene-24-dihydroparkeol; and taraxasterol. The olive oil from the epicarp has lower content of 24-methylene cycloartenol and probably cycloartenol than the virgin oil from the same olives (52). GLC analysis of the triterpenic fraction of olive oil is given in Figure 4.5.

Tacchino and Borgoni (53) determined the saturated straight-chain aliphatic alcohols with even C atoms C_{18} to C_{28} in pressed and in solvent-extracted olive oils. They reported no qualitative but considerable quantitative differences. The total alcohol content was 10-70 (mg/100g) in pressed and 224-434 (mg/100g) in solvent extracted oils. This information may be used for detecting the adulteration (see Chapter 15), of virgin oil with solvent extracted oil.

Tocopherols

Olive oil contains α-tocopherol (Fig. 4.6) in the range of 12-150ppm (15,54-56). Investigators do not agree on the presence of β,γ,δ, tocopherols. Gracian and Arevalo (55) reported that the fluctuation in tocopherol content of olive oil is due to their gradual destruction during processing and storage. Table 4.13 shows the tocopherol content of olive oil and other vegetable oils. It has been observed (57) up to 50% losses of tocopherol and tocotrienol during the refining of olive oil. More tocopherols are lost during deodorization.

Fedeli (1) reported that low acidity olive oils contain small amounts of tocopherols. Olive pomace oils contain higher amounts of tocopherols than the oils from olive pulp (29).

Andrikopoulos *et al.* (58) determined the α and γ-tocopherol content in different types (see Chapter 14) of Greek olive oils by HPLC. They found that virgin olive oil, olive oil, olive pomace oil and refined olive oil contained an average of 113,81,156 and 37 mg/kg of α-tocopherol, respectively. The α-tocopherol level within each type of olive oil appeared to be influenced by different factors. The content of γ-tocopherol averaged 17 and 33 mg/kg of virgin and olive pomace oil respectively. Refined olive oil and olive oil contained very low levels of γ-tocopherol.

Gutfinger and Letan (29) found that lipids extracted from the olive seed kernel were higher in tocopherols than the lipids from the fleshy part of olives (290 mg/g oil vs 186 mg/g). In the former, the relative content of γ-tocopherol was higher (25% vs 7%). Neither olive seed kernel nor olive flesh contained δ-tocopherol. Fedeli (1) reported that the concentration of different tocopherols in olive oil was as follows: 88.5% α-tocopherol, 9.9% β plus γ-tocopherols and 1.6% δ-tocopherol. The tocopherol content of olive oil can be used for detecting adulteration (see Chapter 15) of the oil with seed oils (59). The oxidative stability of olive oil is related to the presence of tocopherols.

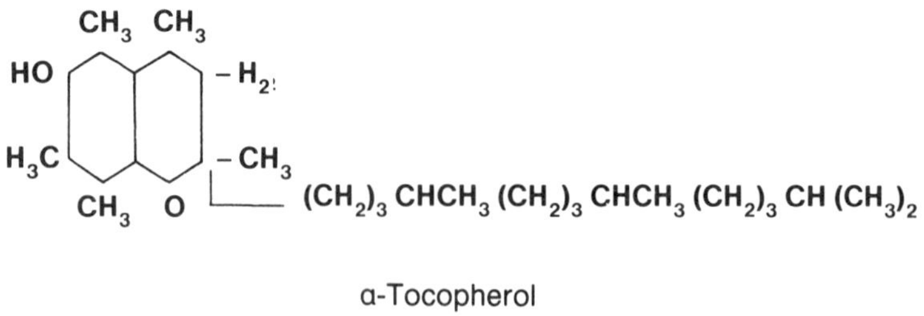

Fig. 4.6. The structure of α-tocopherol

Phenols

The olive mesocarp contains simple as well as complex phenolic compounds which are water-soluble. Small quantities of phenolics, however, are present in the olive oil and increase the oxidation stabilities of olive oils considerably (1,60).

Cantarelli (61) studied the polyphenols in twelve cultivars of olives from central Italy, harvested in two different stages of ripeness, and reported peculiar and differential characters of the examined cultivars. The polyphenol content of olive oil depends on the cultivation procedures, environmental factors, harvesting stage, storage condition of the fruits and extraction procedures (60,62). Sciancalepore (63) reported that the browning rate in five olive cultivars is related to o-diphenol content and polyphenol oxidase activity but not as peroxidase activity.

Graciani *et al.* (64) and Cortesi *et al.* (65) used HPLC for the separation and determination of phenols in olive oil and reported the presence of vanillic, p-hydroxybenzoic and syringic acids. Cortesi *et al.* (65), Vazquez *et al.* (60) and Fedeli (1), found protocatechic, caffeic, p-coumaric, o-coumaric, sinapic, and cinnamic acids in virgin olive oils. Since these acids are absent in refined olive oils, their concentration in oils could serve as an indicator in estimating the presence of virgin oil in mixed oil (65).

Balice and Cera (66) used GC to determine the acidic phenolic fraction of the olives. The juice was extracted with an organic solvent, and the extract was injected into the GC after suitable derivatization. They found p-hydroxyphenylacetic acid and veratric acid in the extract. Vazquez (67) and Solinas and Cichelli (68) used GLC for the separation of phenolic compounds in olive oil and identified tyrosol, hydroxytyrosol and several phenolic acids using gas chromatographic retention times.

TABLE 4.13

Tocopherol content of certain vegetable oils.*

Oils	Tocopherols (mg/g)			
	α	$\beta+\gamma$	δ	total
Olive oil	0.24	traces	traces	0.24
Cottonseed oil	0.56	0.38	traces	0.94
Corn oil	0.26	0.92	traces	1.18
Soybean oil	0.07	0.78	0.24	1.09
Peanut oil	0.23	0.31	traces	0.54

*Gutfinger and Letan (29)

Tyrosol and hydroxytyrosol are derived from the hydrolysis of oleuropein while benzoic and cinnamic acids are derived from the hydrolysis of flavonoids (anthocyanins, flavones), which are found in considerable amounts in the ripe fruit (60,69).

Montedoro *et al.* (62) used TLC to separate the phenolic compounds in the oil during maturation of the fruit and did not notice any quantitative differences between green, semi-black and black olives.

Notte and Romito (70) demonstrated that polyphenols extracted from olive leaves acted as antioxidants in olive oil. These polyphenols were effective in a concentration of more than 20 mg/100g oil. Gutfinger (71) observed that phenols extracted from olive oil with 80% methanol

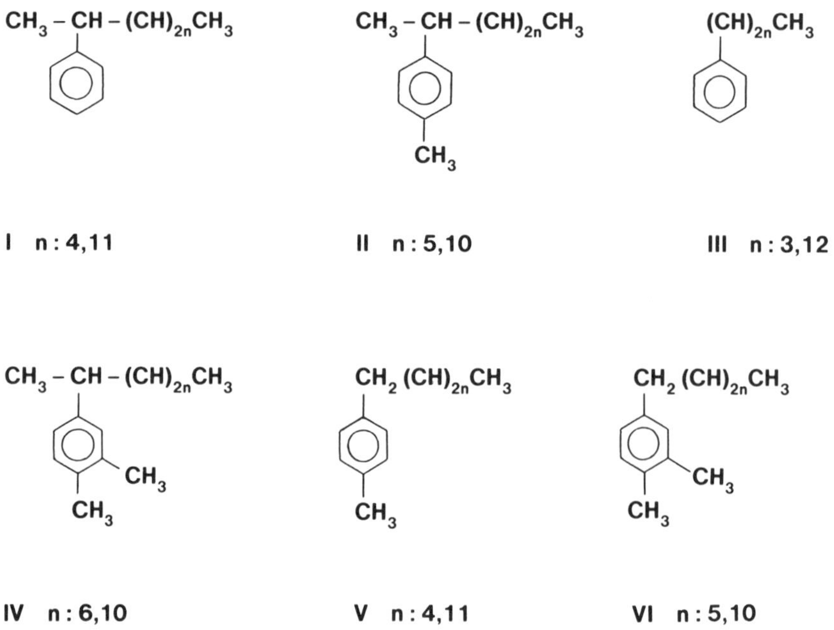

Fig. 4.7. Phenolic compounds in the aqueous phase of the olive fruit.
I. 4-hydroxy-phenyl ethanol
II. 3,4-dihydroxyphenyl ethanol
III.3-hydroxy-phenyl ethanol
IV. 3-hydroxy-methoxy-phenyl ethanol
V. 3,4-dihydroxycinnamic acid (caffeic acid)
VI. 3,4-dihydroxybenzoic acid (protocatechic acid) (1).

acted as antioxidants in other oils while the removal of phenols from olive oil accelerated its oxidation. Olive oil extracted with a chloroform-methanol mixture was more stable to oxidation than mechanically pressed oil and this suggested that stability is due to the higher content of the phenolic compounds in the solvent-extracted oil (71). The amount of water used in the extraction process of olives also affects the concentration of phenolic compounds in the oil. Large quantities of added water remove more phenolics from the olive oil.

The decrease of the phenolic constituents during the mechanical extraction process may be explained by the solubilization effect of the tissue water and particularly by the removal of the colloidal substances (proteins and polysaccharides) which bind these compounds. Figure 4.7 gives the structures of six phenolic compounds found in the tissue water of mature olives. Two of them have carboxylic groups.

Kiritsakis (13) reported an average of 120ppm total phenolic compounds in Greek olive oil which was extracted by centrifugation. Rahmani (72) found that the total phenol and polyphenol content of Moroccan olive oil ranges from 97.5 to 400ppm.

Phospholipids

The main phospholipids of olive oil are phosphatidylcholine, phosphatidylethanolamine, phosphatidylinositol and phosphatidylserine. Their total concentration ranges from 40 to 135ppm (73). Oleic acid is the dominant fatty acid in these phospholipids. More phospholipids are present in olive pomace oil than in olive oil.

Chlorophylls

Olive oil contains chlorophyll a and chlorophyll b (Fig. 4.8) which are green and pheophytin a and pheophytin b which are brown. These pigments are mainly responsible for the unique color of olive oil. Fresh olive oil contains chlorophylls a and b at a concentration of 1 to 10ppm and pheophytins a and b at 0.2 to 24 ppm (1). Chlorophylls a and b are easily degraded to pheophytins. The chlorophyll content of olive oil decreases with the process of fruit maturation. Felice *et al.* (74) found somewhat higher chlorophyll values in the oil obtained by centrifugation than by pressure extraction.

Rahmani (72) reported that pheophytin a was present in amounts ranging from 7-48ppm in oils extracted from green olives, and from 1-13ppm in oils extracted from semi-black olives. Chlorophyll a and pheo-

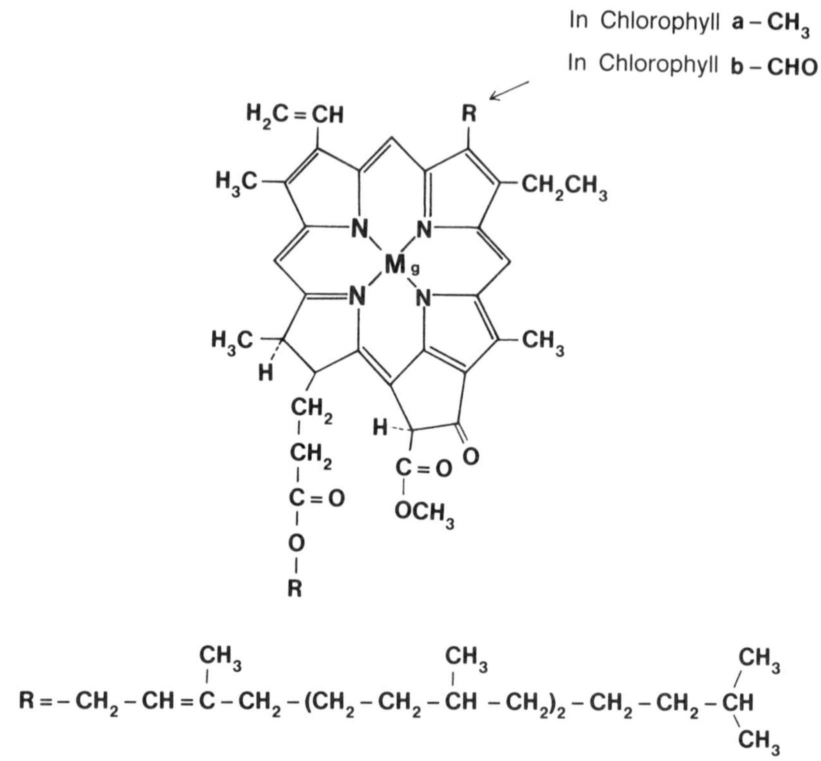

Fig. 4.8. The structure of chlorophylls, a and b.

phytin b were detected only in oils extracted from green olives and the concentrations ranged from trace quantities, to 6-3ppm, respectively. He further reported that pheophytin a is the predominant pigment in Moroccan virgin olive oil. Its concentration in oils is inversely related to the degree of fruit ripening.

Vazquez *et al.* (67) observed that olive oil is very sensitive to 320-700 nm radiation. This sensitivity is due to the presence of chlorophylls which absorb light in that range. Interesse *et al.* (75) reported that the chlorophylls and pheophytins have a prooxidant effect on lipids in the presence of light, while in the dark they act as antioxidants. Vitagliano (54), Kiritsakis (76) and Rahmani (72) also reported that olive oil containing green pigments must be protected during storage from light to minimize the oxidation. These pigments facilitate the formation of singlet oxygen (see Chapter 12) and thereby promote the oxidation of oil in the presence of light (35,36,38,72,77).

Flavor Compounds

The characteristic fragrance and taste of olive oil are due to a number of pleasant flavoring substances (1). The flavor compounds in the olive oils have been studied by several investigators (1,30,62,78-86). Lercker *et al.* (84) found that the concentration of volatile compounds was less than 1ppm in extra virgin oils. They also observed that oils from different regions of Italy differed quantitatively in their volatile constituents. Montedoro *et al.* (62) reported that there are quantitative differences among the volatile constituents of the different cultivars. Fedeli and Jacini (80,81) identified many components which contribute to the olive oil aroma. These include a series of saturated aldehydes ranging from C_7 to C_{12} with C_{10} predominating. Seventy seven volatile components were determined in virgin olive oil.

Evans *et al.* (78) remarked that determination of aroma components by GC may miss easily oxidized substances. Montedoro *et al.* (62) separated and identified volatile compounds of olive oil via gas chromatography mass spectroscopy (GCMS) headspace techniques. Flath *et al.* (1973) used steam to strip the aromatic constituents of California virgin olive oil in a stream of nitrogen. Nawar (30) submitted the oil to high vacuum distillation and used traps cooled by liquid nitrogen to recover the flavoring components. A GC profile of flavor compounds in olive oil and the odor description are given in Figure 4.9.

Fedeli (1) and Montedoro *et al.* (62) classified the aroma compounds of olive oil into aliphatic and aromatic hydrocarbons, aliphatic and triterpenic alcohols, aldehydes, ketones, ethers, esters and furan and thiophene derivatives (Table 4.14). Hexanal, *trans*-2-hexenal, 1-hexanol and 3-methylbutan -1-ol are the major volatile compounds of olive oil (62).

Fedeli (1) identified numerous aromatic hydrocarbons and aldehyde compounds in olive oil (Fig. 4.10) and reported that these compounds are probably formed from the oxidative degradation products of olive oil.

Salib *et al.* (87) identified linalool, linalyl acetate, methyl anthranilate and eugenol in fresh olive oil and camphene, myrcene, menthol, nerol, anethol, methyl acetate and farnesol in aged olive oil.

Colakoglou and Unal (79) reported the presence of the following volatile compounds in olive oil: acetaldehyde, n-propanal, acetone, n-butanal, ethanol, n-pentanal, n-propanol (or methyl-propanol), n-hexa-

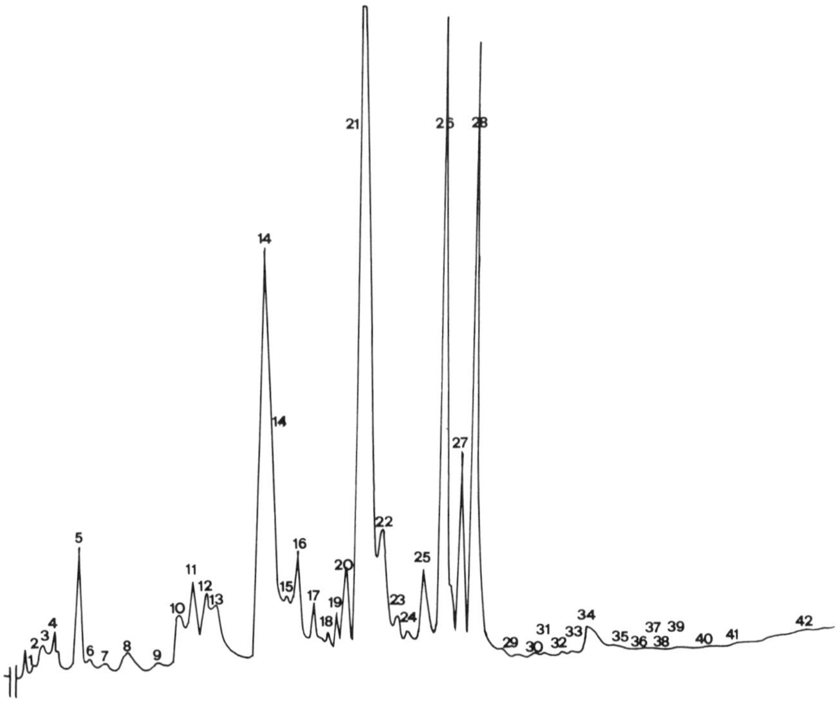

Fig. 4.9. Gas chromatogram of olive oil volatiles (62).

Peak Number	Area %	Odor Description
1	12	Odorless
2	33	"
3	28	Pungent
4	53	Odorless
5	1.52	"
6	26	"
7	14	Ether-like
8	58	Alcohol
9	26	Odorless
10	1.33	Pungent, acrid
11	1.96	" "
12	1.32	Odorless
13	2.33	"
14	12.09	Fruity, leafy
15	1.08	Leafy
16	2.83	Fruity (banana-like)
17	1.35	Leafy, grassy
18	62	Leafy, grassy
19	80	" " olive
20	1.19	Pungent (fusel oil)

21	28.85	Leafy, grassy
22	3.36	Fusel-like
23	1.04	Fruity (olive fruit)
24	99	Pungent
25	2.05	Fruity (banana like)
26	8.45	Fruity (aromatic flavor)
27	2.71	Fruity, grassy
28	8.73	Fatty
29	09	Floreal
30	11	Fruity
31	11	Orange-rose odor
32	16	Almond
33	17	Bread
34	1.23	Odorless
35	33	Fruity
36	14	Terpene-like
37	16	Odorless
38	14	"
39	28	"
40	39	Cooked olive
41	70	Odorless
42	8.92	"

nal, n-butanol (or 2 methyl-butanol) heptanal, n-pentanol and n-hexanol. The hexanal content of the oil increased during 24 months of storage. Rahmani (72) observed that during olive oil oxidation both the saturated and unsaturated carbonyls increased in quantity. He further observed that saturated carbonyls increase much faster than unsaturated carbonyls. Snyder (88) reported that saturated carbonyl compounds, such as pentanal, hexanal, octanal and nonanal are the major compounds formed in oxidized olive oil.

Steam distillation of olive oils and ether extraction of the "juice" showed that the flavor compounds in fresh, aged and rancid olive oils do not contain nitrogen, halogens and sulfur. All these oils contained alcohols, aldehydes, and esters. Salib *et al.* (87) also reported that the infrared spectrum of good quality olive oils is completely different from the spectrum of rancid oils.

Barrio *et al.* (89) applied sensory evaluation and GLC analysis to study the relationship between octane concentration and the flavor quality of olive oil and reported that there is an inverse relationship. As the octane peak area in the headspace increased, the sensory quality

TABLE 4.14

Flavor components of olive oil.*

Hydrocarbons	*Ketones*
Naphthalene	Acetone
Ethylnaphthalene	3-Methylbutane-2-one
Dimethylnaphthalene	Pentane-3-one
Acenaphthene	Hexane-2-one
n-Octane	2-Methyl-2-hepten-6-one
Isopentane	Octane-2-one
2-Methyl-pentane	Nonane-2-one
Ethylbenzene	Acetophenone
Hexane	*Ethers*
Nonane	1,2-Dimethoxybenzene
	Methoxybenzene
Aromatic Hydrocarbons	*Furan derivatives*
	2-Propylfuran (two isomers)
Aliphatic hydroxy compounds	2-n-Pentyl-3-methylfuran
Methanol	2-n Propyldihydrofuran
Ethanol	
Methylpropane-1-ol	
Pentene-1-ol	
3-Methylbutane-1-ol	
2-Methylbutane-1-ol	*Thiophene derivatives*
2-Isopropenylthiophene	2-isopropenylthiophene
Hexane-1-ol	2-Ethyl-5-hexylthiophene
trans-2-Hexene-1-ol	2,5-Diethylthiophene
Heptane-1-ol	2-Ethyl-5-hexyldihydrothiophene
Octane-1-ol	2-Ethyl-5-methyldihydrothiophene
Nonane-1-ol	2-Octyl-5-methylthiophene
2-Phenylethane-1-ol	
Isopropyl alcohol	
	Esters
Terpenic hydroxy compounds	Ethyl acetate
1,8-Cineole	Ethyl propionate
Aldehydes	Ethyl 2-methylbutyrate
n-Propane-1-al	Ethyl 3-methylbutyrate
3-Methylbutane-1-al	1-Propyl 2-methylpropionate
2-Methylbutane-1-al	3-Methyl-1-butylacetate
n-Butane-1-al	2-Methyl-1-propyl
n-Pentane-1-al	2-methylpropionate
trans-2-Pentene-1-al	Methyl hexanoate
Pentene-1-al	cis-3-Hexenyl acetate

TABLE 4.14 continued

Flavor components of olive oil.*

Hydrocarbons	*Ketones*
n-Hexane-1-al	Methyl heptanoate
cis-2-Hexene-1-al	Methyl octanoate
trans-2-Hexene-1-al	Ethyl benzoate
n-Heptane-1-al	Ethyl octanoate
2-4-Hexadiene-1-al	Methyl salicylate
Heptene-1-al	1-Octyl acetate
trans-2-Heptene-1-al	Ethyl phenylacetate
Benzaldehyde	Ethyl nonanoate
n-Octane-1-al	Ethyl decanoate
2,4-Heptadiene-1-al (two isomers)	Ethyl heptanoate
trans-2-Octene-1-al	Ethyl palmitate
n-Nonane-1-al	Methyl oleate
trans-2-Nonene-1-al	Methyl linoleate
2,4 Nonadiene-1-al	Hexyl acetate
trans-2-Decene-1-al	
2,4-Decadiene-al (two isomers)	
trans-2-Undecene-1-al	

*Fedeli (1) Montedoro et al (62)

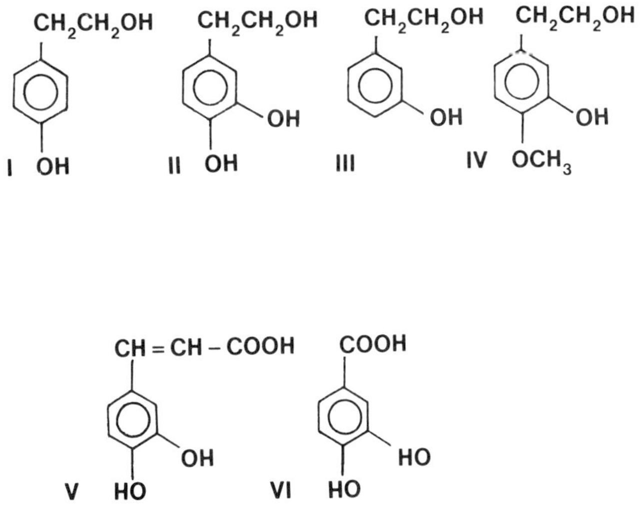

Fig. 4.10. Aromatic hydrocarbons of olive oil (1).

decreased during a storage period of 100 days. The correlation between octane peak area and sensory score was r=0.99. Barrio *et al.* (90) further confirmed their previous observations by analyzing additional numbers of samples from different cultivars, locations and seasons. Highly significant correlations were observed between the octane peak area and the flavor quality of the oils (r=0.964-0.983 for smell, 0.940-0.967 for taste, p=0.01), and demonstrated the existence of a good relationship between flavor quality and octane concentration in the headspace. Gasparoli and Fedeli (91) concentrated the volatile flavor compounds during the GLC analysis of olive oil. The technique applied may be used as an objective instrumental method of olive oil quality evaluation.

Solinas *et al.* (83) studied the relationship between the maturity of the fruits and the flavor components of the oil for the Cultivar "Drittar" using capillary GLC headspace analysis, and observed that no correlation between the fruit pigmentation and the content of 2-hexenal, hexanal, and heptanal, which are major volatile compounds in olive oil. They also concluded that the pigmentation cannot be used for the evaluation of olive oil flavor quality.

Montedoro *et al.* (62) studied the formation of volatile and phenolic constituents of the cultivars Canino, Frantoio, and Moraiolo and found quantitative differences only in the volatile constituents. The patterns of variation and the concentrations of the phenolic constituents in the three cultivars were quite similar. The distribution of the esters, alcohols, and aldehydes in the three cultivars varied similarly over the time of maturation.

TABLE 4.15

Effect of olive storage on the volatile and phenolic constituents of the oil.*

	Oil Extraction	
Volatile Constituents	No storage	After 10 days of storage
Aldehydes %	26.62	13.58
Alcohols %	17.14	25.77
Esters %	4.37	2.24
Hydrocarbons %	11.44	14.14
Others %	40.43	44.27
Total %	100.00	100.00
Total Phenols (mg/kg oil)	104.00	89.00

*Montedoro et al. (62)

Montedoro *et al.* (62) reported that there is some slight change in the total volatile components for the same olive oil cultivar grown in different regions. They suggested that the concentration of the different aromatic compounds in the oil increases with the degree of pigmentation to a certain point during the maturation of the olive. Beyond this point, inversion of this relationship is observed. The highest concentration of the volatile compounds and polyphenol components of olive oil occurs during the period between the semi-black and complete black color of the olives. The maximum concentration of the oil in the fruit is achieved at this stage. Aldehydes are always found in greater quantities compared to other flavor compounds in olive oil. The content of aldehydes is 50% of the total volatile compounds in green olives and 75% in black olives (62).

Formation and Destruction of Flavor Compounds

The formation of volatile compounds is related to cell destruction. An enzymatic process occurs involving hydrolyses and oxidation. The reactions proceed at a high rate depending on pH and temperature.

13-Hydroperoxy-9, 11-octanoic acid is formed by the enzymatic oxidation of linoleate. Subsequently hexenal is formed by the action of the enzyme aldehyde-lyase. Finally, 1-hexanol is formed by an enzyme reduction process. In a different process the hydroperoxyls generate *cis*-3-hexenal and *trans*-2-hexenal by means of aldehyde-lyase. Further reduction leads to the formation of *cis*-3-hexen-1-ol, *trans*-3-hexen-1-ol and 1-hexenol (62).

Vazquez *et al.* (60) reported that tyrosol and hydroxytyrosol which are the major polyphenols of olive oil, are derived from the hydrolysis of oleuropein. Phenolic acids like benzoic and cinnamic probably result from the degradation of flavonoids.

Montedoro *et al.* (62) studied the formation of the volatile compounds of oil in relation to the milling process and found that, when milling and pressing were applied for the oil extraction, more aldehydes were formed than when the fruits were only pressed. They also studied the changes in the volatile compounds of the oil during the storage of olive fruits and reported that the aldehydes and esters decreased during 10 days of storage. Also the total phenolic compounds decreased with storage time (Table 4.15).

Kiritsakis and Min (92) reported that the flavor compounds in olive oil are influenced by many factors. Different cultivars may produce oils of different organoleptic characteristics under identical conditions of environment and cultivation. Climatic and soil conditions, cultivation practices, maturity of fruits and storage conditions also influence the flavor quality of the oil (4). Olives harvested relatively early yield oil with a fruity flavor. Processing techniques (93) and storage conditions (79) also influence the flavor compounds of olive oil.

Sensory Evaluation of Olive Oil Flavor Compounds

Montedoro *et al.* (62) reported that phenolic compounds have limited importance in the flavor of olive oil. Fedeli (93) on the other hand, suggested that the phenolic substances have a significant effect on the olive oil flavor and affect the flavor stability of olive oil.

Among the polyphenols, hydroxytyrosol, tyrosol, caffeic acid, coumaric acid, and p-hydroxybenzoic acid, have the most important sensory characteristics. Montedoro *et al.* (62) reported that hydroxytyrosol is found in very good quality oils, while tyrosol and some phenolic acids are found in oils of poor quality. Some of these compounds have a sharp negative effect when they are present above certain concentrations. This occurs, especially, when hydroxytyrosol is present (62).

Even though much work has been carried out to separate and identify the volatile compounds in olive oil, more research is needed to find out which compounds are specifically responsible for the unique and delicate flavor of good fresh olive oil and how these compounds are formed in olive oil.

References

1. Fedeli, E., Lipids of olives. Prog. Chem. Fats and other Lipids. 15:57 (1977).
2. Fakourelis, N., Effect of pigments on the oxidation stability of olive oil. A Thesis, Ohio State University (1985).
3. Kiritsakis, A. and L.R. Dugan, Studies in photooxidation of olive oil. J. Am. Oil Chem. Soc. 62:892 (1985).
4. Kiritsakis, A. and P. Markakis. Olive oil, A review. Advances in Food Research. 31:453 (1987).
5. Frezzotti, G. and M. Manni. Olive oil processing in rural mills. FAO, Rome, Agricultural development, paper No. 58 (1956).
6. Vitagliano, M., A.M. Leone, and A. Vodret, Determining the purity of olive oil by gas chromatography. Olearia. 14:177 (1960).
7. Amellotti, G., A. Dachetta, D. Grieco, and K. Martin, Analysis of pressed olive oils in Liguria in relation to the olive harvesting period. Riv. Ital. delle Sost. Grasse. 50:350 (1973).
8. Ministry of Commerce. Studies on Greek Olive Oil, (1966,1969,1970,1988).

9. Christakis, G., M.K. Forduce and C.S. Kurtz, The biological aspects of olive oil. Proceedings of the third International Congress on the Biological Value of Olive Oil. Chanea, Greece (1980).
10. Marzouk, B., M. Zarrouk and A. Cherif, Proceedings-World conf. Emerging Technol. Fats. Oils Ind., Baldwin, A.R., Editor. Am. Oil Chem. Soc., Champaign, IL (1986).
11. Gracian, T., The chemistry and analysis of olive oil. In Analysis and Characterization of Oils. Fats and Fat Products. Boekennogen, H.A., Editor. Inter. Publ. a Division of John Wiley & Sons, New York (1986).
12. International Olive Oil Council (IOOC), International trade standards applying to olive oils and olive-residue oils. COI/T.15NC No 1. Madrid, Spain (1984).
13. Kiritsakis, A., Quality studies on olive oil. Ph.D Dissertation. Michigan State University, East Lansing, Michigan, USA (1982).
14. Codex Alimentarius Commission. WHO/FAO, Recommended International Standards for olive oil, virgin and refined and for refined olive residue oil. CAC/RS-33 (1970).
15. Colakoglu, M., Donnees analytiques. Nourelles sur les huiles d'olive. Rev. Franc. des Corps Gras. 13 Annee. 4:261 (1966).
16. Kiritsakis, A., Aspris, P. and P. Markakis, Trans isomerization of certain vegetable oils during frying. In Proceedings of the 6th International Flavor Conference, Recent Developments in Flavor Science and Technology. Rethymno Crete, Greece. Charalampous, G. Editor, Elsevier Amsterdam, N.Y. (1989).
17. Raina, B.C., A.K. Bhatia, S.K. Katiyar and A.K. Gupta, Quality characteristics of promising Himachal olive varieties (OLea europea C.) J. Food Sci. Techn. 23:237 (1986).
18. Rana, M.S. and A.A. Ahmed, Characteristics and composition of Libyan olive oil. J. Am. Oil Chem. Soc. 58:630 (1981).
19. Iverson, T.L. and D. Firestone, Fatty composition of olive oil by urea fractionation liquid chromatography. J. Assoc. Anal. Chem. 48:1191 (1965).
20. Viola, P., International Olive Oil Council Symposium. Advantages Olive Oil. Amman, Jordan (1983).
21. Carbonnier, A. and M. Neuman, Influence de l'huile d' olive sur certaines fonctions hepatiques de l'homme. Int. Congr. Biol. Value Olive Oil. Lucca, Italy (1969).
22. Casares, R., La digestibilidad del aceite de oliva. Int. Congr. Biol. Value Olive Oil. Lucca, Italy (1969).
23. Gyorgy, P., Recent developments on the nutritional role of fats. Int. Congr. Biol. Value Olive Oil. Lucca, Italy (1969).
24. Bailey's Industrial Oil and Fat Products, Swern, D., Editor. Inters. Publ. a Division of John Wiley & Sons. New York (Third edition) (1964).
25. Fedeli, E., The behaviour of olive oil during cooking and frying. In Frying of Food. Varela, G., A.E. Render and I.D. Morton, Editors. Ellis Horwood Ltd.,

Chichester, England (1988).
26. Fedeli, E. and G. Jacini, Lipid composition of vegetable oils. Advances in Lipid Research. 9:335 (1971).
27. Jacini, G., E. Fedeli and A. Lanzani, Research on the nonglyceride substances of vegetable oils. JAOAC 50:84 (1967).
28. Hamann, Y., A. Chaouch and G. Lesgards, Analysis of unsaponifiable contituents of olive oils by HPLC using electrochemical detection. Annal. des Falsif. de l'Exp. Chim. et Toxicol. 81:11 (1988).
29. Gutfinger, J. and A. Letan. Studies of unsaponifiables in several vegetable oils. Lipids. 9:658 (1974).
30. Nawar, W.W., The flavor of olive oil. 44th National meeting of Am. Oil Chem. Soc., Chicago, IL (1970).
31. Cuisa, W. And A. Morgante, Polycyclic aromatic hydrocarbons in olives. Quad. Merced. 13:21 (1974).
32. Stancher, B., F. Zonta and P. Bogoni, Determination of olive oil carotenoids by HPLC. J. Micronutr. Anal. 3:97 (1987).
33. Mehlenbacher, V.C., The analysis of fats and oils. The Gerrard Press Publ. IL (1986).
34. Zonta, F., and B. Stancher, Carotenoids content and quality of olive oils. Riv. Ital. delle Sost. Grasse. 64:53 (1987).
35. Fakourelis, N., E.C. Lee and D. Min, Effects of chlorophyll and β-carotene on the oxidation stability of olive oil. Journal of Food Science. 52:234 (1987).
36. Lee, E.C. and D.B. Min, Quenching mechanism of β-carotene on the chlorophyll sensitized photooxidation of soybean oil. Journal of Food Science. 53:1894 (1988).
37. Frankel, E.N. Chemistry of autoxidation. Mechanism products and flavor significance. In Flavor Chemistry of Fats and Oils. Min D.B. and H.T. Smouse, Editors. Am. Oil Chem. Soc., Champaign, IL (1985).
38. Carlsson, D.J., T. Suprunchunk and H.M. Wiles, Photooxidation of unsaturated oils. Effects of singlet oxygen quenchers. J. Am. Oil Chem. Soc. 53:656 (1976).
39. Matsushita, S. and J. Terao, Singlet oxygen initiated photooxidation of unsaturated fatty acid esters and inhibitory effects of tocopherols and β-carotene. In Autoxidation in Foods and Biological Systems. Michael G. Simic and Marcus Karel, Editors. Plenum Press, New York and London (1987).
40. Boskou, D. and D. Morton, Changes in the sterol composition of olive oil on heating. J. Sci. Food Agric. 26:1149 (1975).
41. Itoh, T., K. Yoshita, T. Yatsu, T. Tamura, and T. Matsumoto, Triterpene alcohols and sterols of Spanish olive oil. J. Am. Chem. Soc. 58:545 (1981).
42. Leone, A.M., E. Notte and F. Lamparelli, The sterols fraction of olive oil and its analytical significance. Riv. Tech. Alim. Nutr. Um. 6:205 (1976).

43. Kiosseoglou, B., I. Vlachopoulou and D. Boskou, Esterified 4-monomethyl- and 4,4-dimethyl-sterols in some vegetable oils. Grasas y Aceites. 38:102 (1987).
44. Boskou, D., Stability of natural terpenoids in heated olive oil. Grasas y Aceites. 29:193 (1978).
45. Meneghetti, O., G. Amelotti and A. Griffini, Identification of sterols in crude and refined vegetable oils by electronic ionization GC-MS. Riv. Ital. delle Sost. Grasse. 64:461 (1987).
46. Tiscornia, E. and G. Betini, Recent analytical data in chemical composition and structure of olive oil. Riv. Ital. delle Sost. Grasse. 49:3 (1972).
47. Lotti, G., R. Izzo and P. Riu, Effects of climate on acid and sterol composition of olive oil. Riv. Soc. Ital. Scien. 11:115 (1982).
48. Paganuzzi, V., Distribution of alcoholic components in the unsaponifiable fraction of olive drupes. IV.4—methyl sterols. Riv. Ital. delle Sost. Grasse. 57:291 (1980).
49. Morchio, G., R. Andreis and E. Fedeli, Total sterols in olive oil and their variation during refining. Riv. Ital. delle Sost. Grasse. 64:185 (1987).
50. Cucurachi, A., L. Camera, F. Angerosa and M. Solinas, Erythrodiol content of dark green olive oil. Riv. Ital. delle Sost. Grasse. 52:266 (1975).
51. Leone, A.M., F. Lamparelli and E. Notte, Sterols and terpene dialcohols in Apulian olive oil. Riv. Ital. delle Sost. Grasse. 55:342 (1978).
52. Paganuzzi, V., Influence of origin and conservation state on the triterpene dialcohols relative content of untreated olive oils. Fette Seifen Anstrich. 84:115 (1981).
53. Tacchino, C.E. and C. Borgoni, Aliphatic alcohols in pressed and solvent extracted olive oil. Riv. Ital. delle Sost. Grasse. 60:575 (1983).
54. Vitagliano, M., Minor constituents of olive oil. Oli Miner. Grasi e Saponi Coloni e Verniki. 37:136 (1960).
55. Gracian, T., G. Arevalo, Los tocoferols gen los aceites vegetables con especial referencial al aceite de oliva. Grasas y Aceites 16:278 (1965).
56. Boatella, R., Analysis of the tocopherols of vegetable oils by gas-phase chromatography. J. Ann. Brom 27:287 (1975).
57. Hernandez Rabascall, N. and J. Riera, Boatella, Variations in the tocopherol and tocotrienol content during refining and hydrogenation of edible oils. Grasas y Aceites. 38:145 (1987).
58. Andrikopoulos, N., M. Hassapidou and A. Manoukas, The tocopherol content of Greek olive oils. J. Sci. Food Agric. 46:503 (1989).
59. Ninnis, L., M. Ninnis, A. Chatoupis and E. Voudouris, Tocopherol content of Greek olive oil and its significance for detection of adulteration. Pract. Acad. Athenon. 37:210 (1969).
60. Vazquez, R., A.J. Del Valle and J.L.M. Del Valle, Phenolic compounds in olive fruits. Polyphenols in olive oils. Grasas y Aceites. 27:185 (1976).
61. Cantarelli, C., Cultivation and pigmentation of the olive. Analytical examination of the polyphenols of the drupes of various cultivars in central

Italy. Natl. Acad. Olives 1st Natl. Congr. on the olives and their products. Spoleto, Italy (In Italy) (1962).
62. Montedoro, G., M. Bertuccioli and F. Anichini, Aroma analysis of virgin olive oil by head space volatiles and extraction techniques. In Flavor of Foods and Beverages. Chemistry and Technology. Charalampous, G. and G. Inglett, Editors. Academic Press, NY (1978).
63. Sciancalepore, V., Enzymatic browning in five olive varieties. J. Food Sci. 50:1194 (1985).
64. Graciani Constante, E., L.C. Colchera Vela and A. Vazquez Roncero, Study of the polar compounds of olive oil by HPLC.I. Adsorption chromatography. Grasas y Aceites. 31:85 (1980).
65. Cortesi, N., A. Ponziani and E. Fedeli, Characterization of virgin and refined oils by HPLC of polar components. Preliminary note. Riv. Ital. delle Sost. Grasse. 58:108 (1981).
66. Balice, V. and O. Cera, Acidic phenolic fraction of the juice of olives determined by a gas chromatographic method. Grasas y Aceites. 35:178 (1984).
67. Vazquez Roncero, A., A study of the polar compounds in olive oil by gas chromatography. Grasas y Aceites. 31:309 (1980).
68. Solinas, M. and A. Cichelli, Determination of phenolic substances in olive oil. Riv. Soc. Ital. Scien. delle Alimen. 10:359 (1981).
69. Montedoro, G. and C. Cantarelli, Phenolic compounds in olive oils. Riv. Ital. delle Sost. Grasse. 46:115 (1961).
70. Notte, E., and N. Romito, Autoxidation of olive oil. Effects of polyphenols. Ind. Agrarie. 9:325 (1971).
71. Gutfinger, J.,Polyphenols in olive oils. J. Am. Oil Chem. Soc. 58:966 (1981).
72. Rahmani, M., Studies of factors effecting olive oil quality with special reference to Morocco. Ph.D Dissertation, Institute Agronomique et Veterinaire Hassan II.Rabat, Morocco (1987).
73. Alter, M., and T. Gutfinger, Phospholipids in several vegetable oils. Riv. Ital. delle Sost. Grasse. 59:14 (1982).
74. Felice, M., T. Gomes and M. Catalano, Oil extraction from olives by continous industrial pressing. 3 years research results. Riv. Ital. Sost. Grasse. 56:361 (1979).
75. Interesse, F.S., P. Ruggiero and M. Vitagliano, Autoxidation of olive oil. Effects of chlorophyll pigments. Ind. Agrarie. 9:318 (1971).
76. Kiritsakis, A., The effect of light on the oxidative deterioration of olive oil. Proc. Int. Congr. Biol. Value Olive Oil. Chanea, Greece (1980).
77. Rawls, H.R. and P.J. Van Santen, A possible role for singlet oxygen in the initiation of fatty acid autoxidation. J. Am. Oil Chem. Soc. 47:121 (1970).
78. Evans, C., D. Helen, A. Moser and G.R. List, Odor and flavor responses to additives in edible oils. J. Am. Oil Chem. Soc. 48:493 (1971).
79. Colakoglu, M. and K. Unal, The changes of some volatile compounds during the storage of olive oil. Proc. Int. Congr. Biol. Value of Olive Oil. Chanea, Greece (1980).

80. Fedeli, E. and G. Jacini, 42nd National meeting of Am. Oil Chem. Soc., New York (1968).
81. Fedeli, E. and G. Jacini, The flavor of olive oil. 44th National meeting of Am. Oil Chem. Soc., Chicago, IL (1970).
82. Flath, R.A., R.R. Forrey and D.G. Guadagni, Aroma components of olive oil. J. Agr. Food Chem. 21:948 (1973).
83. Solinas, M., V. Marsilio and F. Angerosa, Behavior of some components of virgin olive oil flavor in relation to maturity. Riv. Ital. delle Sost. Grasse. 64:475 (1987).
84. Lercker, G., P. Capella and P. Deserti, Volatile and aromatic compounds of extra virgin olive oils. Sci. Techn. Alim. 3:299 (1973).
85. Olias Jimenez, J.M. and Gonzalez-Quijano, R., Gutierrez, Canning of olive oil in commercial type containers. VI. Conservation in glass, tin, polyethylene and polyvinyl chloride containers. Results of the organoleptic tests. Grasas y Aceites 24:79 (1973).
86. Olias Jimenez, J.M., J. Cabtera Martin and R. Gutierrez Gonzalez-Quijano, Sensorial perception ratio of the aroma of packed olive oils. Grasas y Aceites 25:34 (1974).
87. Salib, A.G., G.S. Abdel-Malek, A. Latif and H. Abdel-Gabbar, Chemical composition of odoriferous compounds in Egyptian and Jordanian olive oils. Egyp. Journal of Food Science 4:81 (1976).
88. Snyder, J.M., E.N. Frankel and E. Sclkc, Capillary gas chromatographic analysis of the headspace volatiles from vegetable oils. J. Am. Oil Chem. Soc. 62:1675 (1985).
89. Barrio Perez-Cerezal, A., F. Gutierrez Rosales and R. Gutierrez Gonzalez-Quijano, Application of GLC headspace technique to the olive oil "atrojado" problem. Grasas y Aceites. 32:155 (1981).
90. Barrio, Perez-Cerezal, A., F. Rosales Gutierrer, J. Martin Cabtera, and R. Gonzalez-Quijano Gutierrer, Application of GLC headspace technique to the olive oil "atrojado" problem. Grasas y Aceites. 34:1 (1983).
91. Gasparoli, A. and E. Fedeli, Evaluation of volatile compounds in edible oils: an approach to purge and trap technique. Riv. Ital. delle Sost. Grasse. 64:453 (1987).
92. Kiritsakis, A. and D. Min, Flavor chemistry of olive oil. In Flavor Chemistry of Lipid Foods. Min, D. and T. Smouse, Editors. Am. Oil Chem. Soc. Champaign, IL (1989).
93. Fedeli, E., Flavor chemistry of olive oil. 76th Am. Oil Chem. Soc. Annual meeting. Philadelphia, PA (1985).

Chapter 5

Harvest and Storage of Olive Fruits

The most appropriate stage for harvesting olives destined for oil extraction is when the fruit reaches optimum maturity. At that stage the oil content and the oil quality of the fruit are at their highest level (1,2). Traditionally, olives are harvested at the green-yellow or black-purple stage. Too early and too late harvests have a negative effect on both quality and quantity (2,3). The fruit weight increases and reaches a maximum, before declining along with the volume of the fruit. During senescence, the glycerides commence to hydrolyze and the acidity increases. Since the fruit does not mature at once even in the same tree, the harvesting should take place when the majority of the fruits are at optimum maturity. This is not always possible because other circumstances may affect harvest time, such as

- Availability of farm labor
- Need for early harvest in order to avoid olive fly (Dacus) attack
- Availability of olive oil mills
- Weather, etc.

Harvest Techniques

The methods of harvesting olives remained unchanged for many centuries. Only recently have new techniques been introduced to facilitate the collection of olives. Studies in California showed that 50-70% of the total labor expenses and 30-40% of the gross profit are spent on harvesting.

Harvest after Natural Fall

In the past, olives were harvested by hand after falling on the ground. This method has been largely replaced in Greece and other countries, where farm labor is getting scarce. Plastic nets are spread under the trees and the fruits are let to fall on the nets naturally. This technique is considered very good, especially where the area under the tree is not level and the trees are high. For minimizing the deterioration of the oil, the olives should not be left on the nets more than 15 days (4). Some olive producers, however, leave olives on the nets for longer times and this results in oil of inferior quality. According to Kapatos and Fletcher

(5), the time between fruit drop and harvest is the most important factor in increasing the acidity of the olive oil. This effect is great on the infested fruits and especially the fruits with insect exit holes.

Harvest from the Tree by Hand

Olives to be pickled are almost always picked by hand. This method is also, sometimes, applied to small and young trees, for oil production. Hand picking from the tree, undoubtedly, yields fruit of the best quality for both the table olives and olive oil industries. At the same time, hand picking results in minimum injury to the tree. Jacoboni (6) reported that hand picking, in Italy, covers 60-80% of the total olive cultivation expenses. According to him, no more than 20% of the olive oil produced in Italy originates from olives picked by hand.

Harvest by Beating the Branches

Olives are collected by beating the branches of the trees with long poles, wooden or plastic, after the olive skin has darkened and the oil content has increased substantially. At times plastic combs are used to detach the olives. This method is also costly. In California some oil olive cultivars are harvested with this technique, placing canvas on the ground, then shaking off the fruits from the trees. However, this method sometimes injures fruiting branches and spreads the bacterial disease, olive knot, *(Bacterium savastanoi)* (7).

Great efforts have been made to invent cheap, and efficient methods of harvesting olives such as the Greek made flogging machines. Metzidakis and Michelakis (8) and Siggelakis (9) have studied this type of machinery in detail at the Institute of Subtropical and Olive Tree in Chanea, Greece.

Harvest with Shakers

Mechanically harvested black-type processed fruit seems to be quite acceptable from a quality standpoint. Bark injury has been overcome by properly designed and padded camps (10). Tree shakers and catching frames designed for use with deciduous tree fruits have been used to a limited extent for olive harvesting by a few California growers. Trials with shaking machines in California have shown some promise, but the method has not been generally adapted because of the limited oil olive cultivars and because of the difficulty of soil preparation (10).

In the Mediterranean countries (Italy, Spain and Greece), mechanical harvesting would apply mostly to oil olives, which are picked in late fall or early winter. In California, on the other hand, table olives in the green-to-straw-color stage are the primary target for mechanical harvesting (10).

Difficulty in removing olives by shaking is caused both by a tree structure which is not relatively well adapted to transmission of vibration and a strong attachment between fruit and tree (11). The attachment of olive fruit to trees, as measured by the ratio of removal force to fruit weight (F/W) is very strong compared to that of most other fruits (12).

Abscission chemicals, helping the release of the fruit from the tree have been introduced (13-15). It is expected that the use of abscission sprays and flogging machines together will result in more effective olive harvesting.

Olive Storage

Ideally, the extraction of the oil should follow the harvesting of the olive without delay. This would secure the highest yield and the best quality oil. As this is often impractical, the olives must be stored for a period of time before processing. During storage, several chemical and biochemical changes may occur leading to deterioration of the oil. The most serious damage is caused by fermentation of the olives. This is a term describing the reactions caused by both the enzymes indigenous in the olive tissue and those produced by bacteria, yeasts and molds growing on it. As the olive tissue respires, heat is produced which if not effectively dissipated, will accelerate the enzymatic actions. Lipolysis, lipid oxidation and other undesirable reactions occur.

To avoid olive deterioration, storage in cool buildings is recommended with the olives piled in layers not exceeding 25cm. Spreading the olives on perforated trays and stacking the trays leads to better utilization of storage space.

A common practice is to place the olive fruits in bags made mainly of plastic film. The use of plastic bags should be discouraged, since the conditions created inside them favor microbial growth and spoilage of the fruit (16). Suarez (17) suggested the use of perforated crates, either wooden or plastic which allow ventilation and minimize bruising the fruit. Staking the trays leads to better utilization of storage space. Crates are not widely used except for edible olives.

It has also been suggested to store olives under water in tanks. Adding mild preservatives such as salt (3%), or citric acid (0.03%) and salt (3%), or metabisulfite (2%), to the water would result in better storage (17). Partial drying of the olives with infrared radiation, before storage, has been tried although such a treatment appears too costly. Olive fruit preservation in closed tanks containing inert or antiseptic gases, such as nitrogen or ammonia, has given satisfactory results (17). Olive paste also has been preserved in lined tanks.

These storage methods are costly. An effective and relatively inexpensive way of storing oil olives is to keep them in cool or refrigerated rooms, near the oil mill. Edible olives are often stored in dilute brine until pickling vats for lye treatment become available. Different solutions can be used for the preservation of these olives. Luh and Kean (10) describe the techniques proposed for storing edible olives.

References

1. Papanastassiou, D.P., Modern olive oil processing. The technology of olives. Spiros, Spirou and Son Editors. Athens, Greece (in Greek) (1966).
2. Mikros, L., Olive collection period. A. report. Institute of Subtropical and Olive Trees. Chanea, Greece (in Greek) (1981).
3. Frezzotti, G. and M. Manni, Olive oil processing in rural mills. FAO, Rome. Agricultural development paper No. 58 (1956).
4. Kiritsakis, A. and P. Markakis, Effect of olive collection regimen on olive oil quality. J. Sci. Food Agric. 35:677 (1984).
5. Kapatos, E.T. and B.S. Fletcher, An assessment of components of crop loss due to infestation by *Dacus oleae* in Corfu. Entom. Hellenica. 1:7 (1983).
6. Jacoboni, N., Il problema della raccolta meccanica nella situazione attuale dell' olivicoltura Italiana. In Mecanizzazione della raccolta delle olive. Consigio Nazionale delle Ricerche. Quaderno No 2, April 1978 (1978).
7. Caran, D., Olive harvesting in Turkey. Olivae. 6:19 (1989).
8. Metzidakis, I. and N. Michelakis, Results of trials by olive collecting machines. Olive period 1984-85. Institute of Subtropical and Olive Trees. Chanea, Greece (In Greek) (1985).
9. Siggelakis, C., Application of new olive collecting machines. A report, Eleourgiki. Athens, Greece (In Greek) (1986).
10. Luh, B.S. and C.E. Kean, Canning of fruits. In Commercial fruit processing. Woodroof, J.E. and B.S. Luh, Editors. The AVI publishing company, inc., Westport, Connecticut (1975).
11. Fridley, R.B. and P.A. Adrian, Mechanization and cultural practices in olive production. In Fruit and vegetable harvest mechanization. Cargill, B.F. and G.E. Rossmiller, Editors. Michigan State University, East Lansing Michigan (1969).
12. Lamouria, L.H. and H.L. Brewer, Determining selected bio-engineering properties of olives. Trans. Amer. Soc. Eng. 8:271 (1965).

13. Hartmann, H.T., A. Tambesi and J. Whisler, Promotion of ethylene evolution and fruit abscission in the olive. J. Amer. Soc. Hort. Sci. 95:635 (1970).
14. Hartmann, H.T., W. Reed and K.W. Opitz, Promotion of olive fruit abscission with 2-chloroethyl-tris-o-methoxyethoxy silane. J. Amer. Soc. Hort. Sci. 101:278 (1976).
15. Hartmann, H.T. and K.W. Opitz, Olive production in California. Division of Agricultural Sciences, University of California. Leaflet 2474 (1977).
16. Psyllakis, N., L. Mikros and A. Kiritsakis, Characteristiques qualitatives de L'huile d'olive et les facteurs qui influent sur ces caracteristiques. Proceedings of the third International Congress on the Biological Value of Olive Oil. Chanea, Greece (1980).
17. Suarez Martinez, J.M., Preliminary operations. In Olive Oil Technology. Moreno Martinez, J.M., Editor. FAO, Rome (1975).

Chapter 6

Extraction of Olive Oil

Olive oil is present in the form of minute drops in the vacuoles of the mesocarp cells of the olive fruit (1). Milling (crushing) of the fruits helps in the separation of the oil. During milling the microscopic drops coalesce to form larger drops of oil. The malaxation which follows the milling, leads to the formation of larger oil drops and to the separation of the oil phase (1). Figure 6.1 shows the size of the oil drops after milling of the olives, while Figure 6.2 shows bigger oil drops formed during malaxation.

The milling process affects both the yield and the quality of the oil (2). During this process, exposure of the oil "surface" to the air, should be avoided in order to minimize the loss of aroma and the oxidative deterioration of the oil. Traces of metals picked up from the mills, may cause changes in the sensory characteristics of the oil and may act as catalysts of oil oxidation during storage (see Chapter 12).

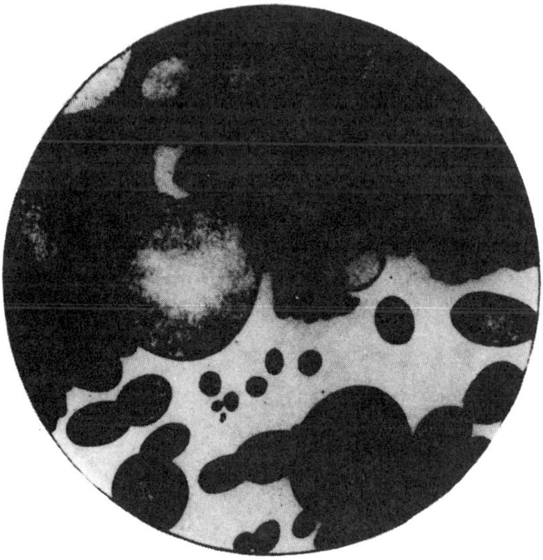

Fig. 6.1. Oil drops after crushing of the olive fruit (2).

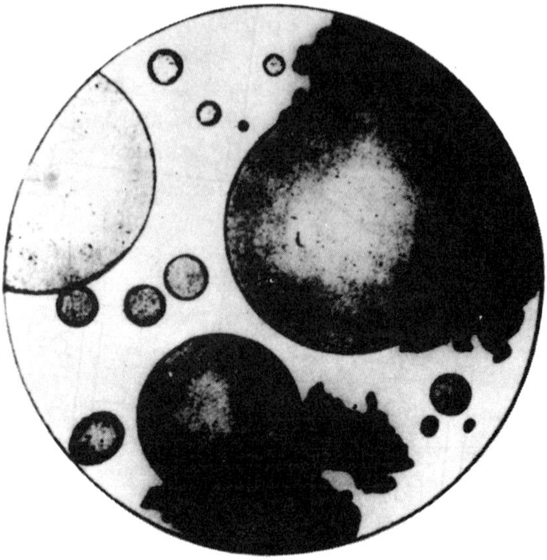

Fig. 6.2. Oil drops after malaxation of the olive paste (2).

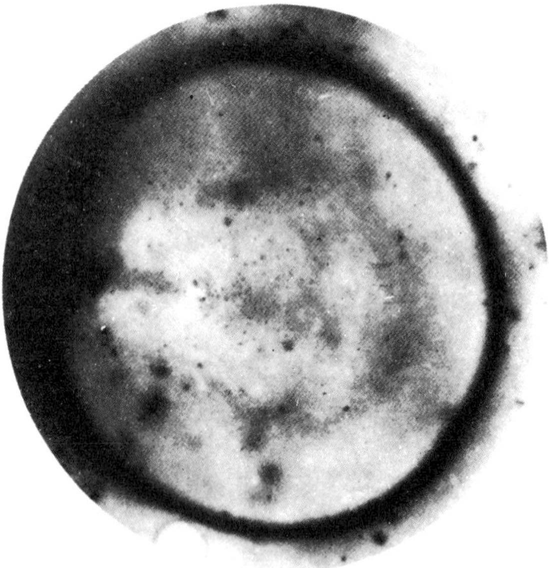

Fig. 6.3. Oil drops covered by lipoprotein membranes (2).

The machinery used and the degree of malaxation play a very important role in the olive oil extraction process. During the milling and malaxation, lipoprotein membranes (Figure 6.3) may be formed on the surface of the original drops, as a consequence of their contact with water. The thickness of these membranes is such as to cause the emulsification of some drops (1). If the malaxation process is slow, the formation of bigger oil drops is possible while at the same time emulsion formation is retarded and the separation of the oil phase is easier. While temperatures decrease oil viscosity and increase oil yields, they should be avoided since they cause biochemical and chemical changes in the olive paste and destroy the sensory characteristics of the oil (2). The increase in the temperature, during the process, may be caused by

- rapid mill rotation
- excessive milling and malaxation, and
- addition of warm water.

When the separation of oil is difficult, the addition of wetting agents may increase the oil yield by 15% to 24%. Figure 6.4 shows the texture of the olive paste before and after the use of wetting substances.

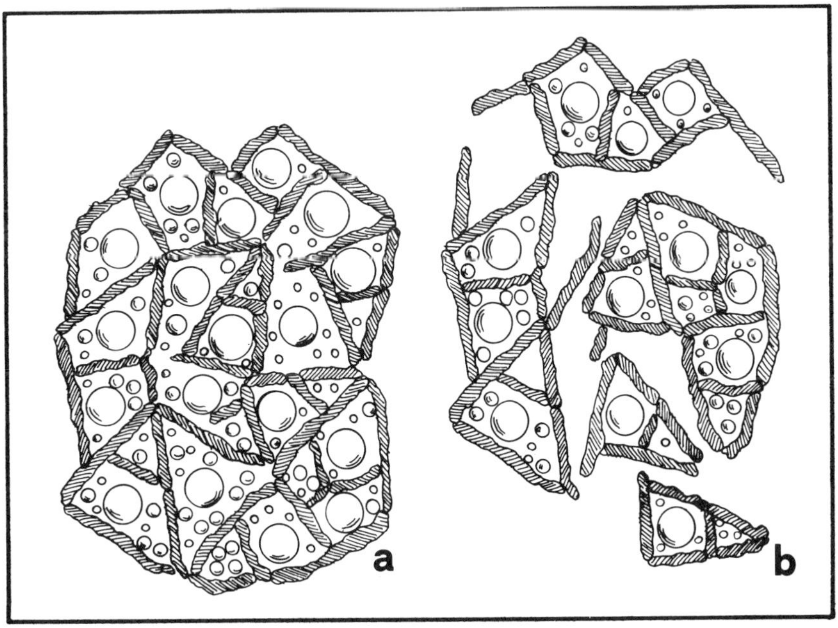

Fig. 6.4. Olive paste before (a) and after (b) the addition of wetting substances.

Processing Steps in the Olive Mill

The main processing steps for olive oil extraction include feeding, washing, crushing, malaxation, separating the olive oil and centrifuging the oil.

Feeding

The fruit is put into a big feeding hopper attached to a moving belt. Removal of the leaves is necessary since they impart a bitter taste to the oil. Passing of chlorophyll from the leaves to the oil also promotes olive oil deterioration (see Chapter 12).

Washing

Washing of the fruit eliminates the presence of foreign materials. A washing machine is a basic accessory of an olive oil processing mill. Suarez (3) experimented with the use of detergents for the washing of olive fruit.

Crushing (Milling)

After being washed, the olive fruit is transferred to the crushing unit. Crushing or milling is the first main step in olive processing. The main types of mills used for olive crushing are described by Mendoza (2).

A milling machine called the edge runner, consists of a large bowl in which two or three heavy wheels revolve, crushing the olives. In ancient mills the bowl and wheels were made of stones. The entire system is surrounded by a metal jacket which prevents the oil from leaking. A piece of metal or wooden equipment directs the olives under the revolving wheels. The revolving of the oil stones is slow and achieves both the crushing and partial malaxation at the same time.

In recent years, newer metal types of olive crushing equipment have been introduced. The metal grinders however may contaminate the oil with traces of metal, unless they are made of stainless steel. Some reasons for replacing of the old crushing mills with new ones include the large size of the old mills, their lower efficiency and the high labor cost.

Malaxation

The malaxation of the olive paste is a very basic step for all the extraction systems. It aids in coalescing of small oil drops into larger ones, thereby facilitating the separation of the oil and water phases.

For better efficiency, malaxators have double walls, or interior tubing for the circulation of the heating water. Increase in the temperature causes changes in the viscosity of the oil and greater olive oil yield (2). However, the temperature of the water should not be high (preferably no more than 30°C) because: the aroma constituents are destroyed, the oil color turns reddish and the acidity is increased.

To avoid the above effects, malaxators must be equipped with an automatic thermostat. The use of steam should be avoided because of its high temperature and for fear of steam leakage.

The malaxators differ in shape and size between olive oil mills. They are mostly cylindrical vats with rotating blades and double walls. A rotating helix with several wings mixes the paste. The velocity of rotation is usually slow (19-20 rpm) and depends on the kind of paste. Prolonged malaxation favors emulsification which retards oil separation (2). For olive fruits that have reached maturity, 20 to 30 min of malaxation is sufficient. Vertical malaxators secure a better air protection for the paste than do horizontal ones. Any part of the malaxator which comes in direct contact with the paste should be made of stainless steel.

Separation of Olive Oil

The main constituents of olive paste are: Olive oil, small pieces of the kernel, water and cellular debris of the crushed olives. Pressure, centrifugation or a combination process may be applied for the separation of oil from the other constituents.

Centrifugation of Olive Oil

A final centrifugation of the oil is needed regardless the process (pressure, centrifugation or combination) used for its separation from the other constituents of the fruit. The centrifuge unit consists of a fixed system and a mobile part which turns very quickly.

A large number of cone-shaped disks are attached to the centrifugal unit. The liquid phase is distributed on the total surface in thin layers and the centrifugation results in a better separation of the oil from the water and other substances. The main factors affecting the separation of olive oil in the centrifugals are:

- "Density", the bigger the difference in density between oil and aqueous phase the easier the separation.

- "Oil drop shape and size", bigger oil drops make the separation easier, while smaller drops result in the formation of emulsions.
- "Viscosity", the thinner the mixture the easier the separation and
- "Temperature", high temperatures aid in the separation of the oil by reducing its viscosity.

Processing Procedures

Pressure Process

A pressure process for oil separation has been used since the beginning of olive tree cultivation. In very old mills both the crushing and the oil separation were carried out with pressure applied either by men or animals (4). The invention of the hydraulic press was a revolution for the function of old type mills and it is used up to date in the improved traditional mills. The olive paste obtained after the crushing and malaxation, is placed in folded cloths made of "esparto grass" or "synthetic fibers". This is accomplished by a loading unit. The next step is to put the folded cloths in a moving unit with a pierced cylinder—guide for the even arrangement of the cloths and support so as to avoid any movement when the system is subjected to hydraulic pressure. A metal tray and a cloth without paste are placed after ever 3 or 4 full cloths for obtaining a uniform application and a more stable load. The moving unit along with its load is placed under a hydraulic pressure unit.

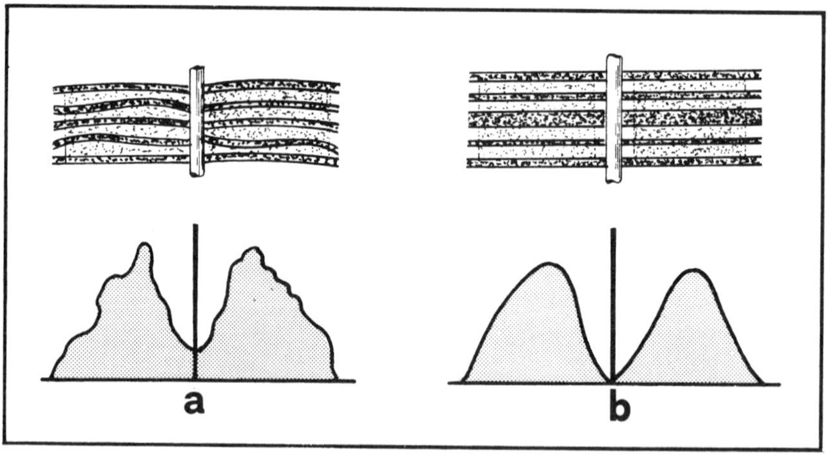

Fig. 6.5. Pressure diagrams in uneven (a) and even (b) distribution of the paste.

Role of Oil "Diaphragms" in the Separation of Olive Oil

Oil diaphragms play an important role, in olive oil processing, because they help to better disperse the paste in thin layers. They further work as filters separating the solid from the liquid phase.

The amount of the olive paste put in the diaphragms (cloths) depends on the type of olive fruits, the duration of milling and the degree of fineness of the paste. An uneven distribution of the paste may cause leakage at the points where the layer is too thin or too thick. Figure 6.5 shows the pressure diagrams when the distribution of the paste is uneven and even. For uneven distribution the pressure is different at different points which causes the diaphragms to wear out more rapidly. For even distribution the pressure applied is higher in the middle and decreases toward the center. The diaphragms used today are mainly made of coco palm fibers or plastic fibers.

Esparto Cloths of Coco-Palm Fibers

Esparto cloths are good filters, but retain a part of the oil and water on the fibers. They must be washed frequently otherwise the remaining oil is subjected to oxidation in the presence of air and affects the quality of the oil passing later. Fermentation takes place in the water resulting in the formation of bad odors which may be absorbed by the oil. Another disadvantage of these diaphragms is that they are not strong and wear out easily.

Plastic Oil Cloths

Plastic oil cloths are not good filters, because the oil obtained may contain small parts of flesh which block the centrifuge and make the oil-water separation difficult. Their advantages are strength and longlife and easy cleaning.

The most desirable characteristics that a diaphragm should have are:

- Good filtering capability
- Relative flexibility to overcome uneven loading
- Small weight for easier handling
- Small thickness
- Small absorbance ability so they will retain the least quantity of oil and water and
- High strength.

Centrifugation Process

Centrifugation is a relatively new process for the separation of oil from the olive paste and it is based on the differences in density between the constituents of the olive paste (olive oil, water and insoluble solids). In centrifugal olive mills the olive paste, after malaxation, is diluted with ample water. After centrifugation in a decanter, it is separated into its three phases: oil, olive cake, and water phase. Figure 6.6 shows the interior of a decanter.

After crushing and malaxation, the olive oil is either completely free, or in the form of small droplets inside microgels or emulsified in the aqueous phase. The separation of the oil, by centrifugation, is much easier when most of the oil is free (2).

The free olive oil can be separated by applying centrifugation while the oil "locked" in the microgels is released by adding a lot of water which changes the coloid constituents. Ranalli (5) found that the oil yield, in a centrifugal process, increased with the degree of comminution of the olive stones. Fine grinding with stone particle size less than 1 mm, 1-2 mm and 2-3 mm reduced the amount of pulp particles lost during centrifugation and increased the yield, without affecting oil quality.

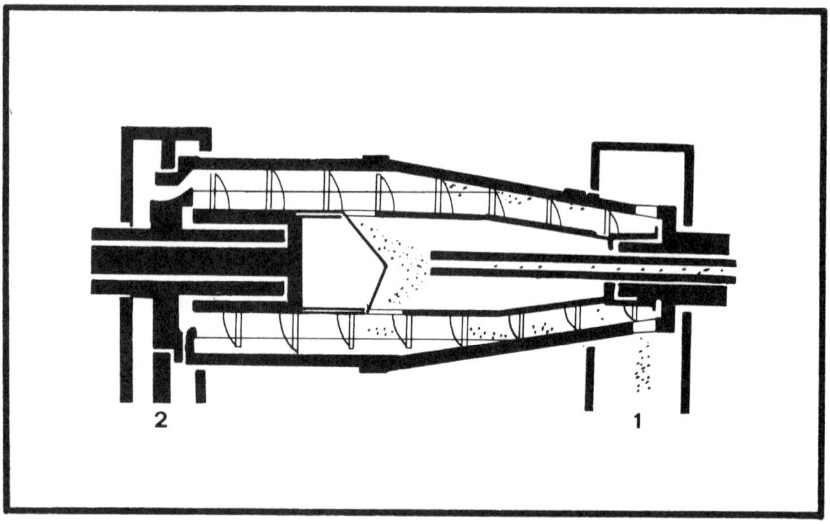

Fig. 6.6. Interior of a centrifuge (decanter). 1. Exit of solid phase. 2 Exit of liquid phase (oil and water separately).

Combination Process

Besides pressure and centrifugation, selective filtration combined with centrifugation is used for the separation of olive oil (see Chapter 7). The most common selective filtration "system" is the sinolea unit. The selective filtration unit separates about 80% of the oil from the olive paste "named" silonea oil. To obtain the remaining oil the olive paste is further moistened, undergoes new malaxation and is subjected to centrifugation. The oil obtained with this process is "called" decanter oil. The water present, in the decanter oil, is separated by a final centrifugation. More details on these processes are given by Petruccioli (6) and Fedeli (7).

Role of Enzymes in the Olive Oil Separation

Enzymes are biocatalysts. Efforts have been made to increase the yield of oil from olives by adding enzymes. Montedoro and Petruccioli (8,9) reported that the enzymes pectin depolymerase, papain, cellulase, hemicellulase, and acid protease increased the oil yield and decreased the extraction time.

Mendoza Alba *et al.* (10) found that the enzymes, pectin-glycosidase, cellulase and hemicellulase (in industrial tests, 300mg/kg) increased olive oil yield by 0.26-11.56%. They did not observe significant changes in the quality characteristics, fatty acid composition, sterol composition and erythrodiol content of the oil. Montedoro *et al.* (11) found that simultaneous use of enzymes and tannin absorbing agents (insoluble polyvinylpyrrolidone, methylcellulose, egg albumin) increased the oil yield and resulted in insignificant modification of the fatty acid composition and of the color of the oil. These enzymes further resulted in a greater stability of the oil and in a qualitative and quantitative increase of its volatile constituents.

Enzymes may be added during the crushing-milling step of the fruit or during the malaxation of the olive paste. Addition of enzymes, during crushing, gives better results (8). It is more common however, to add the enzymes during malaxation. The enzyme olease (a brand name) is recommended for olive oil extraction in cold pressing at a dose rate of 0.05-0.1%. The enzyme disintegrates the intercellular connective tissue of the olive thus increasing the volume of the oil extracted during cold pressing. Research on the effect of enzymes in the oil separation has been done, mainly, in Italy and in olives processed by pressure procedures. Research has been done also in Greece in olives processed by centrifugation procedures.

References

1. Herrera Gomez, C., Mechanical properties of ground olive pastes. In Olive Oil Technology. Moreno Martinez, J.M., Editor. FAO, Rome (1975).
2. Mendoza Alba, J., Milling-Malaxation. In Olive Oil Technology. Moreno Martinez, J.M., Editor. FAO, Rome (1975).
3. Suarez, Martinez, J.M., Preliminary operations. In Olive Oil Technology. Moreno Martinez, J.M., Editor. FAO, Rome (1975).
4. Frezzotti, G. and M. Manni, Olive oil processing in rural mills. FAO, Rome. Agricultural development paper No. 58 (1956).
5. Ranalli, A., Influence of paste stone fragment size on the output of olive oil extraction. Centrifugation system. Riv. Ital. delle Sost. Grasse. 65:13 (1988).
6. Petruccioli, G., Oil extraction. In Olive Oil Technology. Moreno Martinez, J.M., Editor. FAO, Rome (1975).
7. Fedeli, E., Lipids of olives. Prog. Chem. Fats and other Lipids. 15:57 (1977).
8. Montedoro, G. and G. Petruccioli, Enzymic treatments in the mechanical extraction of olive oil. Acta Vitam. et Enzym. 26:171 (1972).
9. Montedoro, G. and G. Petruccioli, Treatment with enzymatic preparation and tannin absorbents in extracting olive oil by single pressure and percolation processes. Riv. Ital. delle Sost. Grasse 51:378 (1974).
10. Mendoza Alba, J., M. Gomez Ruiz, M.C. Gonzalez Prieto and F. Gutierrez Rosales, Efficacy of the enzyme preparation in the technology of olive oil. Organoleptic evaluation of the oil produced. Grasas y Aceites. 38:217 (1987).
11. Montedoro, G., M. Bertuccioli and G. Petruccioli, The effects of treatments with enzymatic preparations and tannin phenolic absorbent agents in extracting olive oil by single pressure process on the oil yield, on the analytical characteristics of the oils, of the vegetation waters, and of the waste waters. Acta Vitam, et Enzym. 30:13 (1976).

Chapter 7

Olive Oil Extraction Systems

Olive oil for many years was extracted from the olive fruits only by pressure type mills (traditional). Recently, new type mills have been developed, such as the ones based on centrifugation or on a combination process.

The oil extraction systems can be divided into those which result in olive pomace of low moisture content (25-30%) and those producing high moisture olive pomace (more than 45%). The traditional olive mills are of the first type, while the second type includes: a) Alfa Laval, Pieralisi, Hiller, Theocharis, Zampekos etc. based on centrifugation and b) Eleourgiki, Rapanelli etc. based on selective filtration. The centrifugal and the combination type olive oil mills are replacing the traditional mills.

Olive Oil Extraction Processes

Pressure Process (Hydraulic Olive Oil Mills)

In this process the liquid and the solid phases are separated by pressure, while the final separation of the olive oil, from the aqueous phase and other substances is accomplished by centrifugation in vertical separators. Moreno Martinez (1) described the pressure as a filtering process. Figure 7.1 shows all of the processing steps for olive oil separation by a pressure type olive mill. Figure 7.2 shows a diagram of a typical pressure olive mill, with three presses working parallel.

Petruccioli (2) and Moreno Martinez (3) reported that several factors affect olive oil extraction in the pressure process. These are: the filtering ability of olive cloths which is related to the presence of olive kernels in the paste, the degree of dispersion of the colloid constituents in the olive paste and their concentration, the water present, the size and the shape of the particles, the physical parameters of the olive oil and the temperature.

Di Giovacchino (4) summarized the advantages and disadvantages of the pressure process as follows:

Advantages:

- Simple, reliable machineries
- Limited investment

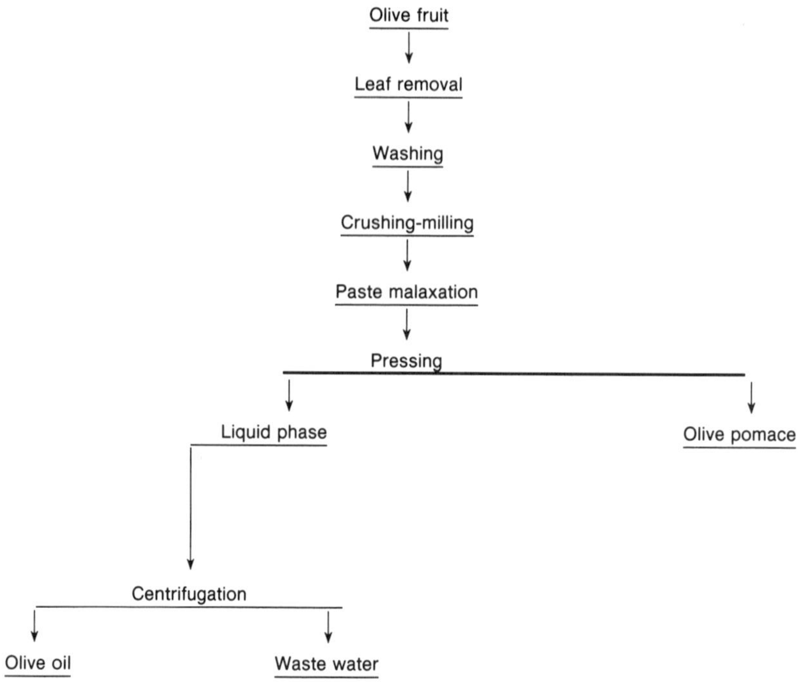

Fig. 7.1. Steps of olive oil extraction based in a pressure process.

- Low water and oil content in the olive pomace
- Small quantity of waste water.

Disadvantages:
- Contamination of the filter cloths
- Labor intensive
- Discontinuous process.

Centrifugation Process

The application of centrifugation for the separation of the oil from the aqueous phase was studied from the beginning of the 20th century. Boulier in 1903, first separated olive oil from the paste, in California, by centrifugation. Several experiments, in different olive oil producing countries, followed. In 1955, the first complete centrifugation olive oil mill for industrial purposes was constructed. Later several types of centrifugal mills appeared internationally.

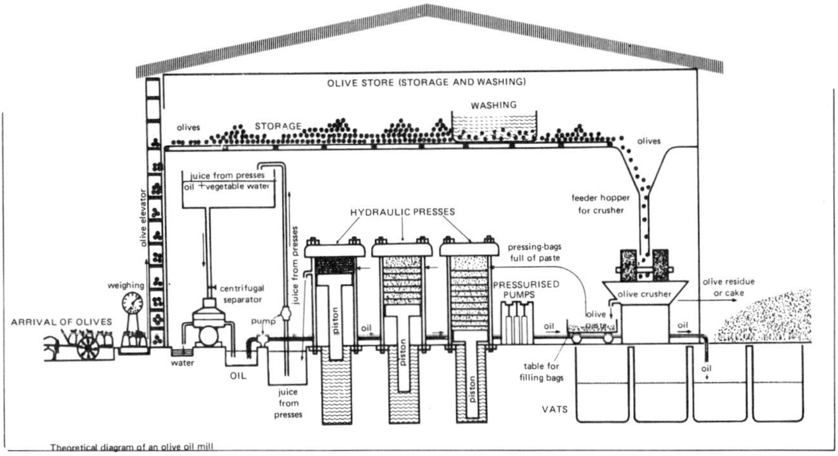

Fig. 7.2. Diagram of a pressure process (Courtesy of IOOC).

The function of the centrifugal olive oil mill is based on the density difference between the constituents of the paste: olive oil, water and olive pomace. Figure 7.3 shows the processing steps in a centrifugation type olive mill.

Centrifugal Olive Oil Mills

In 1965, Alfa-Laval was the first company which made a centrifugal system, the "Centri olive", and in 1969 another one, the "Cosi". In 1971 Pieralisi manufactured a complete centrifugal system. Then many companies (Amenduni, De Vita, Hiller, Theocharis, Zampekos and others) either constructed special decanters for olive oil, or remodeled those used for other purposes and made them suitable for olive oil extraction.

Figure 7.4 shows all parts of a Greek made olive oil processing system with vertical malaxators and vertical centrifuges. This processing unit does not include a second centrifugation. The separation of the paste constituents is accomplished in a vertical centrifuge which is the most basic part of this system.

Combination Process (Selective Filtration-Centrifugation)

Selective filtration was introduced by the Spanish researcher Miguel del Prado de Acapulco who observed that a thin layer of olive paste free of

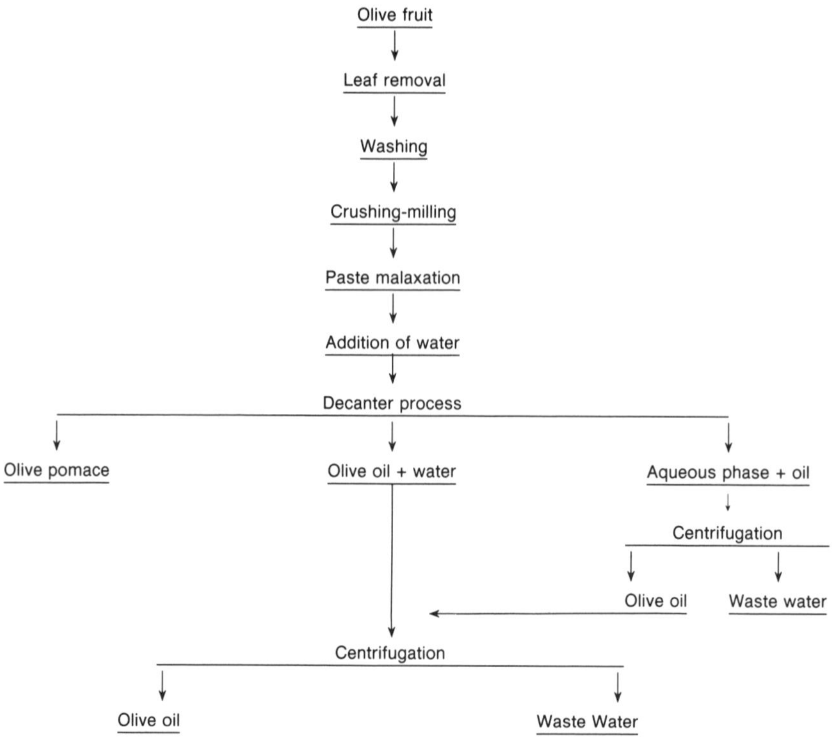

Fig. 7.3. Steps of olive oil extraction based in a centrifugation process.

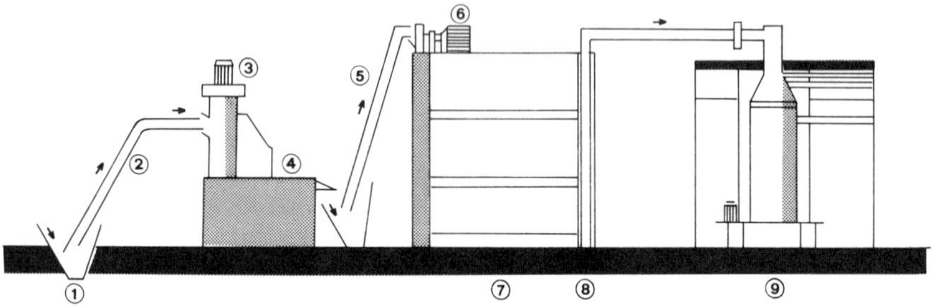

Fig. 7.4. Diagram of a centrifugal olive mill with perpendicular centrifuge (Courtesy of the Zampekos Co.). 1. Feeding of olive fruit 2. Lifting of olive fruit 3. Removal of leaves 4. Washing of fruit 5. Lifting of washed fruit 6. Crushing 7. Malaxation 8. Olive paste pump 9. Centrifuge (perpendicular).

kernels, passed through cotton, produced olive oil free from water. This observation initiated a new extraction system based on selective filtration. The filter initially used, was made of cotton but it was soon realized that another material more lasting and easier to clean should be used (2,3).

In the fifties, a French company manufactured a system based on the Acapulco system with a considerable improvement in the filter and the malaxators. This system considerably decreased the time needed for the oil separation. A more significant improvement, in the initial Acapulco system, was made by Buendia who made the Alfin type extractor. The basic difference between the Alfin and Acapulco systems was mainly in the shape and the kind of the filter used.

The filter of the improved selective filtration unit consists of about 1500 stainless steel metal sheets 12x2 mm (2,3). These metal sheets are attached to a very slow moving base. As the system is moving, the numerous blades get into the paste and when they are withdrawn they take along small oil drops. Figure 7.5 shows how the Alfin type extractor works.

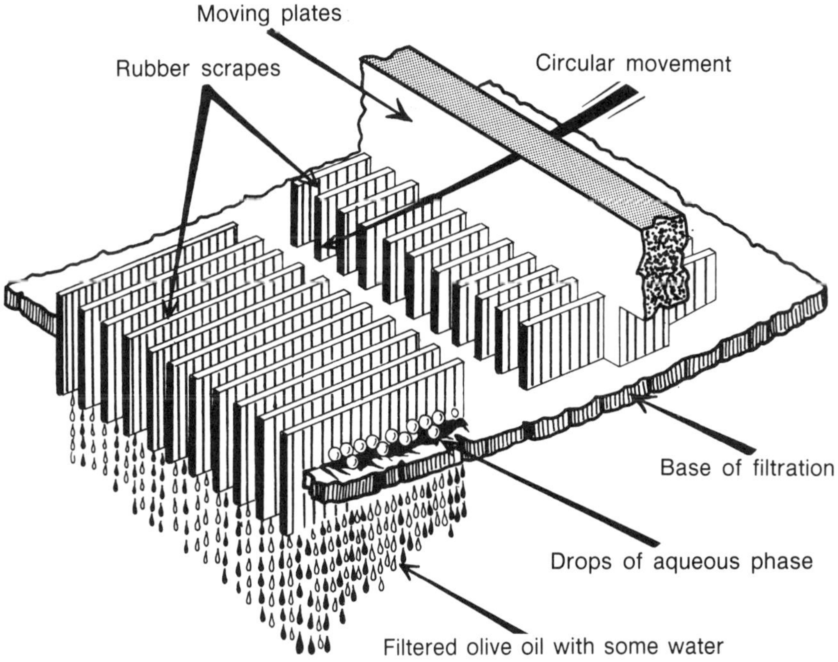

Fig. 7.5. Diagram of the Alfin filtering system.

The Alfin extractors were superior to Acapulco system. More specifically the filter was more durable, it had a larger area of contact with the olive paste and it was easier to clean.

Combination type Olive Oil Mills

Rapanelli in 1972 was the first to introduce a processing system based on both selective filtration and centrifugation. The main parts of this system are the selective filtration unit, known in the industry as sinolea, and the decanter which is a type of horizontal centrifuge (5). Most of the oil contained in the olive fruit (70-80%) is separated with the sinolea unit

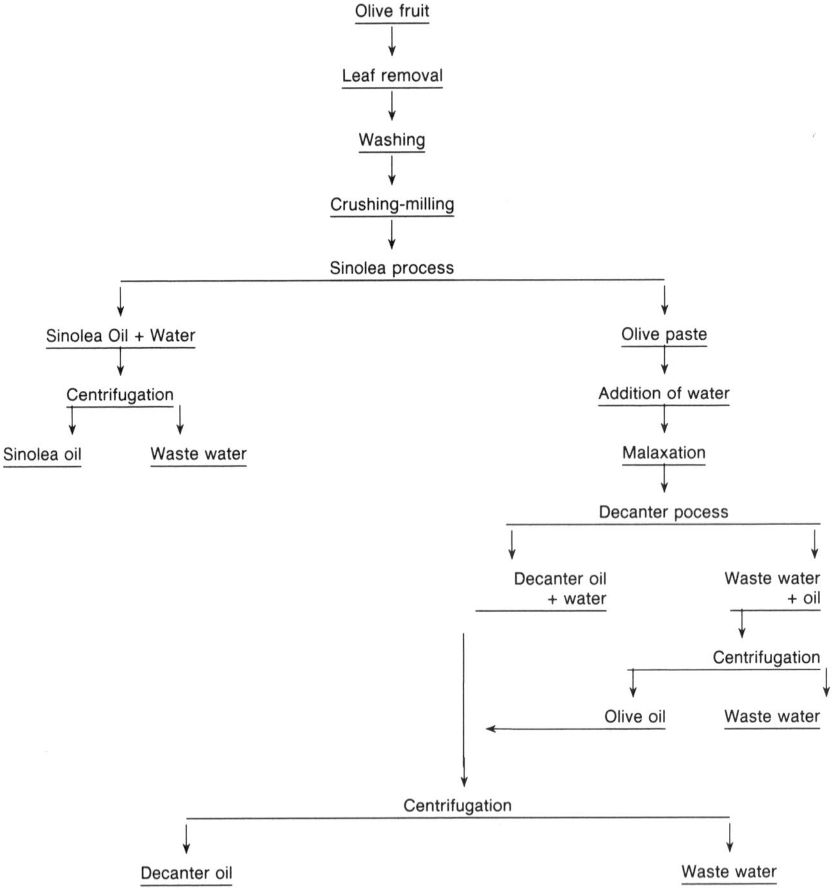

Fig. 7.6. Steps of olive oil extraction based in a combination process.

and taken to the centrifuges for the water and other substances to be removed. The remaining 20-30% of the oil is gained after dilution of the paste with hot water, malaxation and centrifugation in a decanter (decanter oil). This oil is subjected to a second centrifugation for removal of the remaining water.

Figure 7.6 shows all the processing steps in a combination olive oil mill (sinolea-decanter). Since the sinolea oil is obtained without adding hot water, its aroma and quality are very high. The decanter oil has a darker color (contains more chlorophyll) than the sinolea oil because it undergoes further malaxation (5,6). Combination type processing systems of different efficiency and potential, are made in Greece only by the Cooperative Organization "Eleourgiki". The main parts of such a processing system are shown in Figure 7.7.

The main advantages of the centrifugal and combination olive oil processing systems are the following:

- A drastic decrease in hand labor, since almost everything is done mechanically
- Oil with lower acidity is obtained
- Better preservation of the volatile constituents in the sinolea oil (5).

The main disadvantages of these processing systems can be summarized as follows:

- They are expensive
- The oil contains phenols in low concentration due to the losses by the addition of water.

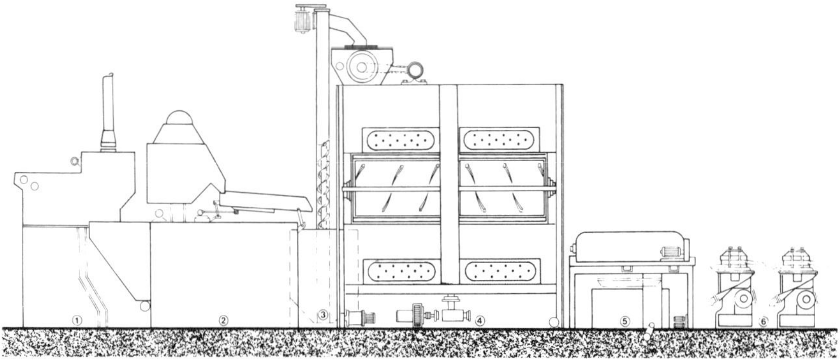

Fig. 7.7. Diagram of the main steps of the Eleourgiki processing mill. 1. Removal of leaves 2. Washing of the fruit 3. Crushing 4. Malaxation 5. Separation (Decanter process) 6. Centrifugation.

The noted advantages are only achieved with equipment made by specialized manufacturers. The olive oil processing systems must be carefully tested for the following:

- Degree of automation
- Hourly potential
- Industrial output
- Quality of the metal used
- Ability to check the temperature at the different stages of processing
- Presence of foreign substances in the oil
- Loss of oil in the olive pomace and in the waste water
- Overall quality of the oil obtained (7-10).

Fedeli (11), Frezzotti and Manni (12), Petruccioli (2) and Kiritsakis *et al.* (13) describe these processing systems.

Laboratory Olive Oil Mills and Apparatus

Small experimental mills have been invented for the determination of the oil content of the olive fruit. Their operation is mainly based on centrifugation. Once the fruit is crushed, the malaxation of the paste takes place in small malaxators. Centrifugation and separation of the solid and liquid phase follow. The olive oil is obtained after a second centrifugation. The determination of the olive oil content can be accomplished by the Soxhlet method. With a new Soxhlet apparatus less time for the extraction of the oil is needed and the solvent is regained. The oil content of the fruit can also be determined by instruments based on infrared absorption or reflection.

References

1. Moreno Martinez, J.M., C. Herrena Gomez and J. Pereda, Estudios fisico-quimicos sobre las pastas de aceitunans molidas. XXI. La prensada de la pasta como un proceso de filtration. Grasas y Aceites. 15:299 (1964).
2. Petruccioli, G., Oil extraction. In Olive Technology. Moreno Martinez, J.M., Editor. FAO, Rome (1975).
3. Moreno Martinez, J.M., Other systems of extraction. In Olive Oil Technology. Moreno Martinez, J.M., Editor. FAO, Rome (1975).
4. Di Giovacchino, L., Olive processing systems. Separation of the oil from the must. Olivae. 26:21 (1989).
5. Carocci, C., Ancora complesso oleario "SINOLEA"per 1′ estrazione dell' olici della olive. Perl l' Olive l' Oleif Imperia (1963).
6. Koutsaftakis, A., N. Psyllakis, A. Kiritsakis, L. Micros and D. Vamvoukas, Comparative trials of olive oil extraction systems. Period 1978-79. Insti-

tute of Subtropical and Olive Trees, Chanea, Greece (In Greek) (1979).
7. Koutsaftakis, A., N. Psyllakis, N. Michelakis and D. Vamvoukas, Results of the experimental olive oil mill. Period 1974-76. Agricultural Research station. Chanea, Greece (In Greek) (1976).
8. Koutsaftakis, A., L. Micros, N. Psyllakis and D. Vamvoukas, Comparative trials, of Decanters. Period 1977-78. Institute of Subtropical and Olive Trees, Chanea, Greece (In Greek) (1978).
9. Koutsaftakis, A., N. Michelakis, D. Vamvoukas and S. Vekiaris, Comparative trials of olive oil extraction systems. Period 1984-85. Ibid (1985).
10. Koutsaftakis, A., I. Metzidakis, D. Vamvoukas and E. Stefanoudakis, Results of centrifugal extraction system (Kalis type L-120). Period 1985-86. Ibid (1986).
11. Fedeli, E., Lipids of olives. Prog. Chem. Fats and other Lipids. 15:57 (1977).
12. Frezzotti, G. and M. Manni, Olive oil processing in rural mills. FAO, Rome. Agricultural development paper No. 58. (1956).
13. Kiritsakis, A., A. Koutsaftakis and L. Mikros, The effect of processing systems "PIERALISI,HILLER and RAPANELLLI (Sinolea-Decanter)" on olive oil quality. Grassas y Acietes. 35:165 (1985).

Chapter 8

Subproducts of the Olive Oil Processing

Olive pomace and waste water are the main subproducts of olive oil processing. The olive tree leaves, which "accompany" the fruit may also be considered as subproducts, because they can be used for soil enrichment. Pomace is the pulpy material remaining after the extraction of the oil from the fruit.

Olive Pomace

Olive pomace contains fragments of skin, pulp and kernel. The quality of olive pomace varies according to the type of olives and the method of processing. It ranges from 25 to 50% of the weight of the olives processed by pressing (1).

The commercial value of the olive pomace depends on its oil and water content which in turn depends on the type of process applied (pressure, centrifugation, combination), and the operating conditions. Most of the olive pomace is subjected to further processing to obtain olive pomace oil. The pomace oil is derived mostly from the olive kernel, especially the endosperm, while part of it comes from the pulp. It has quantitative and qualitative composition similar to olive oil.

Olive Pomace Composition

Olive pomace contains proteins, cellulose and other constituents. Table 8.1 shows the main composition of the olive pomace.

TABLE 8.1

Composition of olive pomace obtained from a press olive oil mill.*

Constituents	Content (%)
Water	25.0
Nitrogen compounds	4.4
Non-nitrogen extractable compounds	20.0
Cellulose	40.0
Ash	6.6

*Carola (2)

The moisture of olive pomace obtained by the pressure process is low (1), while the centrifugation process results in high moisture content (2,3).

Deterioration of the Olive Pomace

The oil present in the pomace undergoes rapid deterioration due to the moisture present which favors the triglyceride hydrolysis. Carola (2) reported that the acidity of pomace oil may be increased from 5% to 60% in a short time. This is due to the presence of the enzyme lipase originating either from the olive fruit or from microorganisms (Gliomastix chartarum, Cephalosporum Sp., Aspergillus glaucus etc.), grown on the pomace. These microorganisms are favored by high moisture and temperature, during storage. Besides the increase in the acidity, oxidation products (aldehydes and ketones) are formed, during storage of the pomace which affect, significantly, the quality of the pomace oil (2,4).

Although much experimental work has been done to find a proper method of storage in order to minimize the deterioration of the pomace oil, the problem still exists. The best approach is to process the olive pomace as soon as possible.

Extraction of the Olive Pomace Oil

Pomace must be dried for the extraction of the pomace oil. Drying of olive pomace is accomplished in long revolving horizontal cylinders through which hot air is passing. "Kernel-wood" (olive pomace after the oil is extracted) is used for heating the air. The olive pomace, hot and dry, is taken to the extractors where it is mixed with an ample amount of solvent. The solvent mainly used is hexane. The amount of hexane used affects the quantity of oil extracted.

The extraction product (a solution of oil and solvent) is called "miscella". The miscella is filtered and then subjected to distillation to remove solvent from the oil.

This oil, called "olive pomace oil" contains large amounts of FFA which increase the acidity and lower its quality (see Chapter 12). It also contains pigments (eg. chlorophylls, anthocyanins) as well as other constituents contributing a characteristic flavor and sweet taste. The pomace oil is inferior to virgin olive oil. Small amounts of refined pomace oil can be added to virgin olive oil (see Chapter 14).

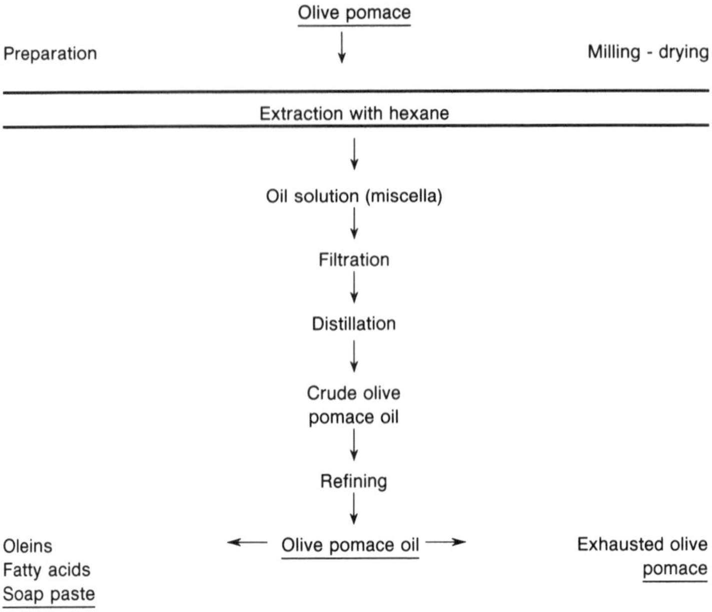

Fig. 8.1. Treatments and subproducts obtained during olive pomace processing.

Refining of Pomace Olive Oil

The olive-pomace oil can be consumed only after being refined. This includes neutralization, deodorization, and decoloration (see Chapter 9). The products obtained from refining pomace oil are: the refined pomace oil, FFA, soap paste and exhausted olive pomace. Treatment of the soap paste, with sulphuric acid, results in release of the fatty acids and in the formation of oleins. Figure 8.1 shows schematically the products obtained during olive pomace processing.

Exhausted Olive Pomace (Kernel wood)

Exhausted olive pomace, known as "kernel wood", is the subproduct of the olive processing. It contains a large amount of fiber, cellulose and some proteins (Table 8.2).

It is mainly used for heating water in the olive oil mills in a special boiler. Small kernel particles can be used for the production of plastic and other materials. The ash of kernel wood is a good fertilizer due to its P, K, and Ca, content (1).

TABLE 8.2

Average composition of the kernel wood.*

Constituents	Content (%)
Water	6.9
Proteins	6.0
Fats	1.0
Cellulose	26.4
Ash	6.5
Extractable material	53.2

*Carola (2)

TABLE 8.3

Average composition of waste water.*

Constituents	Content (%)
Water	83.2
Inorganic constituents	1.8
Organic constituents	15.0
-Lipids	(1.0)
-Nitrogen compounds	(2.0)
-Sugars	(7.5)
-Organic acids	(1.5)
-Polyalcohols	(1.5)
-Pentoses, tannins	(1.5)
-Glycosides	traces

*Carola (2)

Waste Water

The olive fruit may contain more than 50 percent of water. This water along with the water added during the processing steps in the olive oil mill ends up as waste water.

Waste water is rich in sugars. It also contains organic acids, small amounts of emulsified olive oil, minerals and some other constituents. Its average composition is given in Table 8.3. About 20% of the inorganic salts (mostly carbonates and silicates) present, are water insoluble. Fe, Ca, P, and K compounds are the main water soluble salts of waste water (2).

Phenols, polyphenols and some bitter substances, especially oleuropein are also found in waste water. The waste water constitutes either float at the surface or are dissolved in it. Its pH is about 5.5. Waste water is black and has an unpleasant odor when it undergoes fermentation. It is a polluting material. Efforts have been made to build units for the biological control and reuse of waste water. Vigo et al. (5) proposed electrolytic treatment of waste water and Medici et al. (6) suggested anaerobic fermentation of it.

Uses of Waste Water

Waste water has been used for the growth of the yeast *Torulopsis utilis*. Considerable amounts of yeast are produced in Spain by this method (2). The yeast is used in the stock feed industry. Because the composition of waste water is rich in salts only the addition of ammonium sulfate is needed.

Codounis et al. (7) isolated certain anthocyanins from waste water. Waehner et al. (8) produced butanol from waste water in a yield from 0.09-0.29 g/g sugar content, using the microorganism *Clostridium sp.* Waste water can be used as an agricultural fertilizer after treatment with Ca (0.5 kg/100lt), due to its high content of K and P.

Paredes et al. (9) observed an enhanced microbial activity in soil to which waste water was added, but the salinity of the soil was also increased. It is desirable for the waste water of olive oil mills to be converted into some usable product and not add to environmental pollution.

References

1. Frezzotti, G. and M. Manni, Olive oil processing in rural mills. FAO, Rome. Agricultural development paper No. 58 (1956).
2. Carola, C., By products. In Olive Oil Technology. Moreno Martinez, J.M., Editor. FAO, Rome (1985).
3. Koutsaftakis, A., N. Psyllakis, A. Kiritsakis, L. Mikros and D. Vamvoukas, Comparative trials of olive oil mills. Result-report for the period 1978-79. Institute of Subtropical and Olive Trees, Chanea, Greece (In Greek).
4. Fedeli, E., Lipids of olives. Prog. Chem. Fats and other Lipids. 15:57 (1977).
5. Vigo, F., L. Avalle and M. De Paz, Smaltimento delle acque di vegetazione provenienti da frantoi di olive. Riv. Ital. delle Sost. Grasse. 60:125 (1983).
6. Medici, F., C. Merli and E. Spagnoli, Anaerobic digestion of olive mill waste waters. A new process. Anaerobic Dig. Proc. Int. Symp. 4th. Shanghai, Peop. Rep. China, (1985).
7. Codounis, M., K. Katsaboxakis and D. Papanicolaou, Progress in the extraction and purification of anthocyanin pigments from the effluents of olive oil extracting plants. Prog. Food Eng. 7:567 (1983).

8. Waehner, R.S., R.A. Mendez and A.M. Giulietti, Olive black water as raw material for butanol production. Biolog. Wastes. 23:215 (1988).
9. Paredes, M.J., E. Moreno, A. Ramos Cormenzana and J. Martinez, Characteristics of soil after pollution with waste waters from olive oil extraction plants. Chemosphere. 16:1557 (1987).

Chapter 9

Chemical and Other Treatments of Olive Oil

Olive oil of good quality is characterized by a fragrant and delicate flavor which is appreciated by the international gourmet and cherished by the native consumer. This olive oil is almost unique among vegetable oils that can be consumed without any refining treatment, while other vegetable oils must undergo several chemical treatment before becoming edible.

Olive oil needs chemical or physical treatment when it is not of good quality, which occurs when it is extracted from poor quality olives and subsequently maltreated. The quantities of olive oil to be refined are determined by such factors as: seasonal conditions, pest infestation, ripeness (1), harvest practices and storage conditions of the fruit (1,2). The amount of olive oil that is being refined has been decreased recently because:

- The collection methods have been improved
- The storage facilities have also been improved
- The time between collection of fruit and extraction of the oil has decreased and
- Most of the olive oil mills have been replaced with modern, improved mills.

The olive oil that finally undergoes refining is subjected to some or all of the following treatments:

- Removal of resinous substances
- Neutralization of FFA
- Deodorization and
- Decolorization (Bleaching).

Olive pomace oil undergoes a similar refining treatment (see Chapter 8)

Removal of Resinous Substances

This process results in the removal of any mucilage and resinous substances present in the oil. These substances are oil soluble only when they are anhydrous. They become insoluble, with the addition of water, and separable with centrifugation or by a filtering process. Because hydration is accomplished during the neutralization of FFA with alkali solution, the two treatments can be applied simultaneously.

Removal of Free Fatty Acids (Neutralization)

There are two alternatives for removing free fatty acids. One is the neutralization with alkalies and the removal of the soap formed.

$$\underset{\text{Fatty acid}}{R-COOH} \quad + \quad \underset{\text{Sodium hydroxide}}{NaOH} \rightleftarrows \underset{\text{Soap}}{R-COONa} \quad + \quad \underset{\text{Water}}{H_2O}$$

The soap is removed by settling or centrifugation, followed by several washings.

And the second is glycerine esterification. Glycerine is esterified with the FFA present in the olive oil. This technique can be used with olive oil regardless of its acidity. Neutralization with alkalies becomes very costly when the acidity is less than 15-20%. The esterification reaction is:

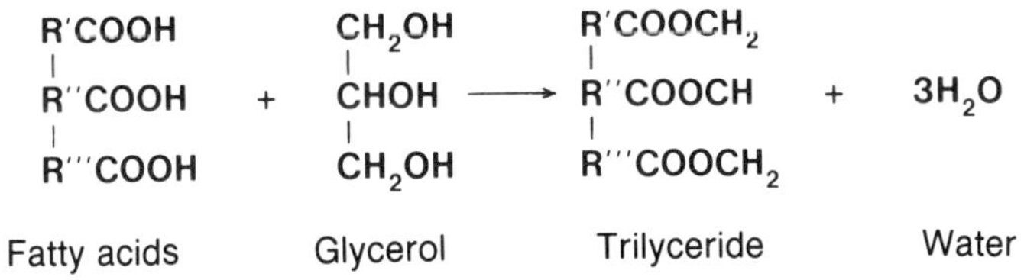

During neutralization several constituents of the olive oil, like tocopherols, are destroyed. Thus tocopherols are often added (260 mg/kg) to refined olive oils for that reason.

The amount of oil lost during neutralization is calculated using the formula:

$$\text{Loss (\%)} = \frac{\text{Weight of crude olive oil - weight of refined olive oil}}{\text{weight of crude olive oil}} \times 100$$

The following formula can be used for the determination of the alkali needed for neutralization (3):

$$X = \frac{B \times O}{A} = \frac{40 \times O}{282} = \frac{O}{7}$$

where,
X = quantity of NaOH needed in g
B = molecular weight of NaOH (40)
O = olive oil acidity expressed in g of oleic acid/100 g olive oil
A = molecular weight of oleic acid (282)

Deodorization

Deodorization is the process that removes the substances (mostly aldehydes and ketones), responsible for the undesirable odor and taste. This process occurs at low pressure (3-5 mm Hg) and high temperature (180-230°C). Deodorization is usually applied to old olive oils with bad odors due to oxidation (see Chapter 12). Fresh olive oil very seldom needs to be deodorized. Residues of pesticides are also removed with deodorization. Six g of vapor (115°C) /hr over 100 g of olive oil are enough for total removal of the pesticides: lindane, aldrin, dieldrin, heptachlor and hexachlorocyclohexane.

Decolorization (Bleaching)

Decolorization removes mostly natural pigments (chlorophylls, pheophytins) present in the oil. Two common means of olive oil decolorization are activated carbon (1-2%) and absorbent clays. The decolorization treatment is applied under vacuum and mild heating (less than 100°C). For successful decolorization the acidity of the oil must not exceed 3% (expressed in oleic acid).

Decanting (Racking-Off)

Good quality olive oils must be decanted periodically to remove solid particles and water that settle at the bottom of the containers forming a blackish layer, commonly known as oil "sludge". These settled substances may undergo fermentation and impart off-flavors to the oil. A disadvantage of decanting is that the olive oil is enriched with oxygen which causes oxidation (see Chapter 12).

Filtration

Most of the particles are removed by decantation. The rest must be removed by filtration. When the oil is not naturally limpid, it has to be filtered in order to remove hydrated colloids present in it. Filtering must

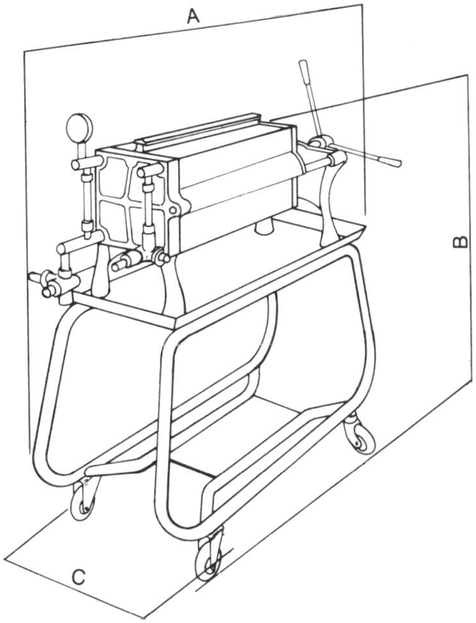

Fig. 9.1. Press for filtering olive oil, with pump (Courtesty of the Ocim Co.).

be carried out just before packing. This process entails expenses and loss of oil.

Filter presses using cotton fabric are used in filtering olive oil (Fig. 9.1). These are fitted with a pump which forces the oil to pass through a number of filters. The particles that pass through the filters settle with time forming a "sludge".

Winterization

Olive oil as well as olive pomace oil are winterized. This treatment removes glycerides of high melting point and thereby avoids clouding and solidification of oil when it is stored at low temperatures.

Winterization is usually applied to olive oil used for the production of margarine and mayonnaise. Solidification of the triglycerides, during storage of margarine and mayonnaise at low temperature, could possibly cause the decomposition of the emulsion of these products.

In the winterization the oil is cooled gradually to temperatures as low as 2°C. During this process, which may last 36 hours, high melting point glycerides crystallize and are removed by filtering. Faster cooling is avoided because it results in small crystals which are difficult to filter out. Solvent winterization methods have also been applied resulting in higher yields and lower manpower costs (1).

Hydrogenation

Hydrogenation is a chemical treatment changing unsaturated fatty acids to saturated ones. This results in higher melting glycerides. Olive oil is mixed with hydrogen under specific conditions of temperature and pressure in the presence of a catalyst. Ni is one of the catalysts used. The catalyst is removed by filtering after the hydrogenation is completed. The hydrogenation process is checked occasionally, by measuring the iodine value, and can be discontinued at any desired level. If the hydrogenation reaction is stopped before completion, a soft solid fat is formed. This type of fat is used in the manufacturing of margarine. More advanced hydrogenation results in fats used in shortening manufacture. The composition of the hydrogenized olive oil depends on the hydrogenation conditions, and on the initial fatty acid present in the oil. Hydrogenation causes decolorization of the olive oil since the pigments (chlorophyll and pheophytin) are destroyed.

References

1. Fedeli, E., Lipids of olives. Prog. Chem. Fats and other Lipids 15:57 (1977).
2. Kiritsakis, A. and P. Markakis, Effect of olive collection regimen on olive oil quality. J. Sci. Food Agric. 35:677 (1984).
3. Papanastassiou, D.P., Modern type olive mills. Olive oil technology. Spiros-Spirou and Son., Editors. Athens, Greece (in Greek) (1966).

Chapter 10

Olive Oil Storage

Olive oil is subject to deterioration during storage (1). The degree of deterioration depends on the storage conditions and is a continuous and irreversible process due to oxidation (see Chapter 12). Besides oxidation, fermentation occurs in small particles that are not removed by filtering. These particles contain sugars which support the fermentation.

Estimation of the Resistance of Olive Oil to Oxidation

Proper storage extends the marketing "life" of olive oil. When olive oil is stored for a long time and in large amounts, the determination of its resistance to oxidation, by certain accelerated tests (Oven, AOM, Rancimat and others), is necessary.

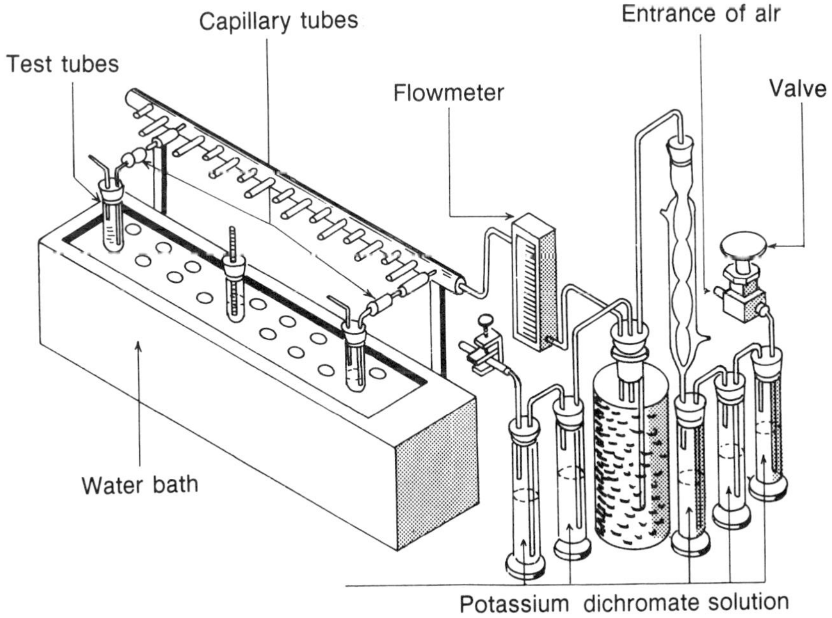

Fig. 10.1. Active oxygen method (AOM) apparatus

The accelerated tests are useful only for comparing the storage stability of different oils and they cannot predict the exact time at which a given oil will deteriorate. These tests can also be used for the evaluation of antioxidants. Based on the accelerated tests, olive oils of different stability should be stored separately and those with higher susceptibility should be marketed earlier. Olive oils extracted from different olive cultivars should also be stored separately.

Oven Test

An olive oil sample of 50g is placed in an oven of constant temperature (usually 63°C) and the oxidation process is followed by chemical and sensory tests. The time, in days, needed for the oxidation to start (see Chapter 12) shows the resistance of the olive oil to this deterioration. The size of the sample container and the amount of the oil tested affect the oxidation rate. A small quantity of oil as a thin layer will predict a shorter stability period than a large sample tested as a thicker layer (2).

Active Oxygen Method

The active oxygen method (AOM) is also sometimes known as the "Swift" test. The olive oil is placed in test tubes set in a water bath at 97.8°C. Air is bubbled through the samples (Figure 10.1). This is a very quick method. The oxidation process is monitored by chemical or sensory tests. Results are expressed in hours.

The resistance to oxidation is very different under the condition of these two tests (Oven, AOM). Identical samples may show a 100 hour

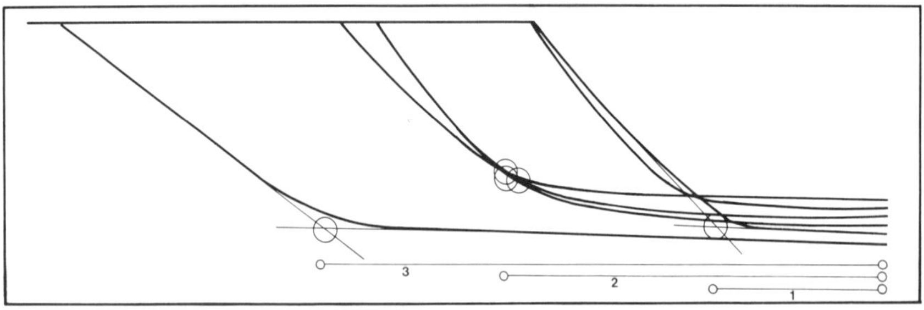

Fig. 10.2. Automatic recordings of oxidation (by Rancimat apparatus) of three different olive oil samples.

resistance to oxidation by the AOM, and a 75 day resistance by the oven test. Each test shows relative stability of samples under conditions of that test.

Other Procedures for Determining the Olive Oil Resistance to Oxidation-Rancimat Procedure

Another method of determining the resistance of olive oil and other lipids to oxidation is the Warburg procedure, which measures the consumption of oxygen by a sample under specified conditions. Another procedure is based on the conductivity change of the distillate collected from an oil sample subjected to accelerated oxidation. The distillate contains formic and other acids which affect its conductivity (3). The Rancimat apparatus is based on this procedure. Figure 10.2 shows recordings, of the initiation and propagation steps, obtained with the Rancimat apparatus. This apparatus also can be used for the determination of the effectiveness of antioxidants in a given sample of oil.

Storage Equipments

Olive oil should be stored in cool places, particularly when storage lasts several months. The containers can be either fixed or movable. In winter, it is advisable to maintain a temperature of 10 to 15°C. At this temperature oxidation is eliminated and clouding is avoided (1). Underground tanks are recommended for the storage of oil in large quantities. Thus the oil is protected from large temperature fluctuations. The material used for construction of these tanks should be physically and chemically inert to the oil in order to avoid absorption of defective odors and flavors. At times metallic drums are used for the storage of olive oil. Such drums may impart a metallic flavor to the oil (as soaps are formed) and promote the oxidative deterioration of the oil (see Chapter 12). Lining the drums and tanks with epoxy resins improves their inertness. Enameled tiles and glass are considered the best lining material for large storage tanks (1).

Cerezal et al. (4) found that olive oil stored up to 325 days in a tank coated with epoxy resin was much more stable than olive oil stored in an iron tank with respect to an FFA content, peroxide value (PV), metal content and sensory characteristics. Table 10.1 shows that olive oil stored in metallic drums without a lining had a higher iron content, higher UV absorbance (170mm) and undesirable sensory characteristics, compared to olive oil stored in lined drums. The lower PV of the olive

TABLE 10.1

Deterioration of olive oil stored in lined and unlined drums.*

Quality changes					
Treatment	Peroxide value	1% E 270nm 1 cm	Loss in Tocoph. and other reduc. mater. (%)	Sensory score (1-10)	Iron Content (ppm)
Control (unlined drums)	5.0	0.584	47	1.6	87.0
Lined drums	11.8	0.193	14	7.2 (acceptable)	2.8

*Cucurachi (1)

oil stored in unlined metal drums does not mean that this oil was less oxidized than the other one, but probably the peroxides formed were broken into secondary carbonylic compounds. The use of a stainless steel container for storing large quantity of olive oil is ideal.

Attempts are still being made to find the ideal lining for olive oil tanks since they affect considerably the quality of olive oil during storage. In a general way, tanks or drums used for storage of olive oil should meet the following specifications:

- Be constructed of material impermeable to oil so they can be thoroughly cleaned before they receive new oil
- Be inert so that they do not react with the oil, absorb odors or contain metals accelerating the oxidation
- Protect the oil from light and air
- Maintain a practically constant temperature, preferable near 10°C.
- Be resistant to impact and pressure, easily manipulated (opened, closed) and
- Be economical.

References

1. Cucurachi, A., Final operations. In Olive Oil Technology. Moreno Martinez, J.M., Editor. FAO, Rome (1975).
2. Kiritsakis, A., C.M. Stine, and L.R. Dugan, The effect of selected antioxi-

dants on the stability of virgin olive oil. M. Sc., Thesis, Michigan State University, U.S.A.

3. Laubli, M.W., and Bruttel, P.A., Determination of the oxidative stability of fats and oils. Comparison between the active oxygen method (AOCS cd.12-57) and the rancimat method. J. Am. Oil Chem. Soc. 6:792 (1986).
4. Cerezal, A., Del Bario Perez, Gutierrez Rosales, R. Gutierrez Consalez-Quijano, Virgin olive oil conservation in storage tanks coated with epoxy resins. Grasas y Aceites 26:287 (1975).

Chapter 11

Packing of Olive Oil

Proper packing of the olive oil is very important for the shelf life and marketing of the product. The most common containers are tin plated or plastic bottles. Polyvinylchloride (PVC) and polyethylene are materials widely used. Glass bottles and lined cartons are found in the international market. Glass is the material with the greatest inertness.

Table 11.1 lists the main "types" of containers used for packing olive oil, and their characteristics. Packaging materials must exclude light and atmospheric oxygen to prevent deterioration of the oil. The elimination of these factors is not always sufficient to conserve unaffected the sensory characteristics of the oil.

Figure 11.1 shows the progress of olive oil oxidation stored in glass and plastic bottles. Table 11.2 and Figure 11.2 also, show the oxidation of olive oil in plastic bottles, exposed to diffused room light.

Unal (2) stored virgin olive oil in cans, colorless glass bottles and PVC bottles, in light (ambient temperature) and in darkness (at 28 + 1°C) for

TABLE 11.1

Main "types" of containers for packing olive oil.*

	Characteristics			
Container	Impermeability to fat	Impermeability to gases	Protection against light	Resistance to impact
Tinplate	E	E	E	Sg
Special board	Sg	G	E	A
Colourless glass	E	E	B	B
Topaz glass	E	E	A	B
Light glass	E	E	B/A	B
PVC	E	G	B	Sg
Polypropylene	Sg	Sg	A	G
Polyethylene[1]	A	A	B/A	G
Polyethylene[2]	Sg	Sg	A	G

*Cucurachi (1)
[1]Low-density polymer
[2]High-density polymer
E: Excellent; G: Good; Sg: Semi-good; A: Average; B: Bad

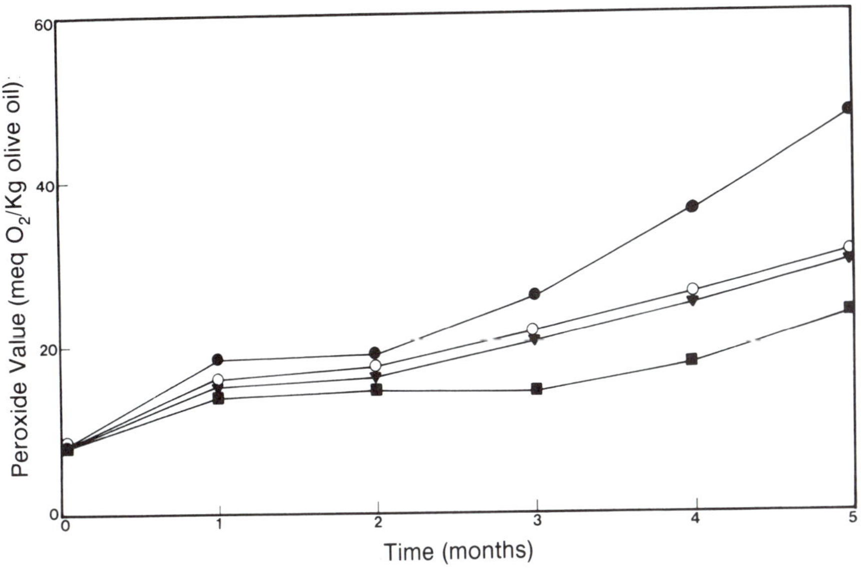

Fig. 11.1. Effect of diffused light on peroxide formation in olive oil stored in plastic or glass bottles. (●: plastic bottles; ▲: plastic bottles wrapped in aluminum foil; ○: glass bottles; ■: glass bottled wrapped in aluminum foil.)

24 months, and noted an increase in FFA in all of them. He also observed that the PV of samples in cans or glass bottles decreased during storage, due to the peroxide decomposition, whereas those of PVC bottles increased, probably because of the O_2 permeability of PVC. He further demonstrated that the decrease in tocopherols, β-carotene and chlorophylls was greater in illuminated samples than in those stored in darkness. Samples stored in glass or PVC bottles, in the light, underwent greater changes in sensory characteristics than those stored in darkness (3).

Gutfinger *et al.* (4) studied the changes in PV, TBA and total acidity in olive oil samples stored at 30°C in the dark, in glass, PVC bottles, and tin-coated cans. They found no significant differences between PV of the oil stored in glass and PVC. Lower PV were determined for oil stored in cans for 160 days.

Cucurachi (1) reported that the oxidation of olive oil in open containers, exposed to unlimited air, was high. On the other hand the formation

TABLE 11.2

Effect of packing container on olive oil deterioration (Storage period three months).*

Olive Oil	Container	Storage conditions	Quality Changes: Peroxide value	Quality Changes: Loss in Tocoph. and other reduc. materials (%)	Sensory differ.
Virgin (fine)	Tinplate	Dark	<20	8	Very significant
	Lined board			15	Very significant
	Glass			4	Very significant
	PVC			4	Significant
	Polypropylene			10	Not significant
	High-density polyethylene			13	Not significant
	Low-density polyethylene			18	Significant
	Glass	Light	<20	35	Very significant
	PVC			30	Very significant
	Polypropylene		20-40	50	Very significant
	High-density polyethylene		30-80	50	Very significant
	Low-density polyethylene		15-45	60	Very significant

TABLE 11.2 continued

Effect of packing container on olive oil deterioration (Storage period three months).*

Olive Oil	Container	Storage conditions	Quality	Changes	Sensory differ.
			Peroxide value	Loss in Tocoph. and other reduc. materials (%)	
Riviera	Tinplate	Dark	<20	10	Not significant
	Lined board			15	Significant
	Glass			20	Not significant
	PVC			18	Not significant
	Polypropylene			20	Not significant
	High-density polyethylene			25	Not significant
	Low-density polyethylene			50	Not significant
	Glass	Light	<20	35	Very significant
	PVC			35	Very significant
	Polypropylene		40-60	50	Very significant
	High-density polyethylene		40-60	55	Very significant
	Low-density polyethylene		20-35	75	Very significant

*Cucurachi (1)

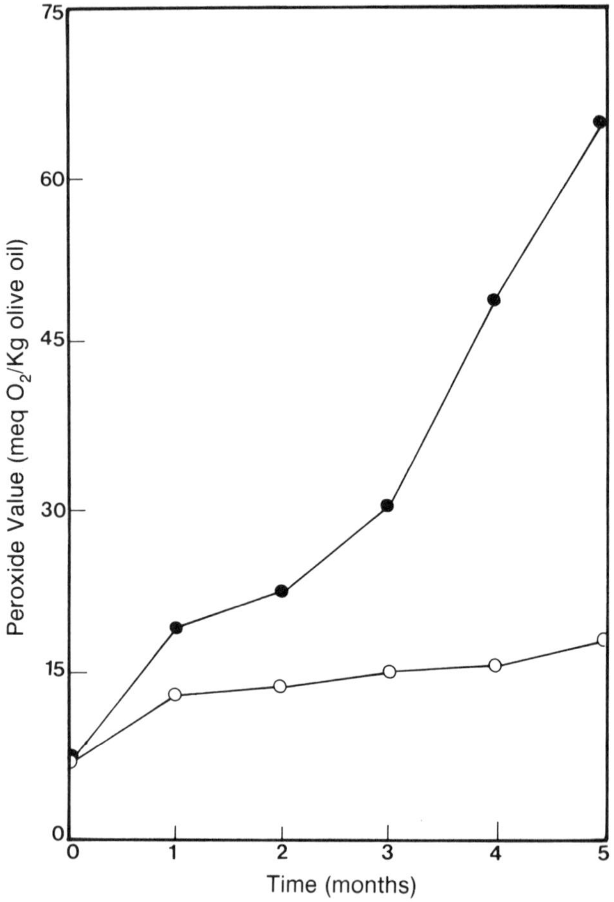

Fig. 11.2. Effect of diffused light on peroxide formation in olive oil stored in plastic bottles (●: plastic bottles; ○ plastic bottles wrapped in aluminum foil).

of peroxides in closed tins, due to the limited amount of oxygen, present in the headspace, was insufficient to produce the typical rancid odor. According to him olive oil samples with the same PV had dissimilar sensory scores when they were packed in different materials.

Kiritsakis (5) studied the relative effect of plastic and glass bottles on the oxidation of olive oil exposed to diffused light. Samples in polyethylene bottles developed higher PV than those in glass bottles. Samples in bottles wrapped in aluminum foil displayed lower PV than those uncovered. Oil stored for 5 months in glass bottles wrapped in aluminum foil had PV lower than oil stored in plastic bottles also wrapped in aluminum foil.

In a general way, the containers used for packing olive oil should meet the following specifications:

- Be sealed in a manner that renders tampering impossible
- Protect from the effect of oxygen, metals and light
- Be resistant to impact and pressure, but be easily held and opened
- Should leave a minimum volume of headspace (Packing under vacuum or inert gas has been suggested (6)
- Should have nondrip opening and other convenient features and
- Be economical.

The aforementioned suggestions are at times overlooked for reasons of cost and attractiveness.

The European Economic Community (EEC) subsidizes the packaging of olive oil in small containers, which are more easily marketed and result in diminished spoilage. The 5-liter and 1-liter containers are favored in this subsidy.

Strangely enough EEC is not yet against the use of clear plastic bottles, which is recommended by others. This author strongly believes that the use of opaque bottles is a necessity. A small area of the container should allow viewing of the oil.

When small olive producers have oil extracted in cooperative or private olive oil mills, the oil should be stored properly in central facilities, from which the producers can later withdraw desirable quantities.

International Rules for Olive Oil Packing

According to the rules and regulations set by the International Olive Oil Council (IOOC) in 1985, the olive oil and the olive pomace oil to enter international trade must be packed in containers of a certain kind and type such as:

- Tanks or tankers or wooden barrels
- Metallic sealed drums, lined with the proper type of varnish (epoxy resins)
- Tin plated cans or plastic bottles, and
- Small and large glass bottles.

Container Filling Tolerance

The olive oil should never occupy less than 90% of the volume. In the case of tin plated cans of 1 liter or smaller volume, the minimum filling is 80%.

Labeling of the Containers

Packed olive oil should be labeled as follows: *Name of product.* The name of the product should be mentioned as well as the specific type (eg extra olive oil, see Chapter 14). When the olive oil has a commercial name (eg Riviera), both names should be written with letters of the same size (eg olive oil Riviera). *FFA (Acidity).* Acidity should be expressed either as percentage of oleic acid or as degree of acidity (see Chapter 12). *Net content.* Labels should mention the net content in weight or volume, expressed International Metric System, units. *Name and Address.* The name and the address of the packing company, the distributor, the importer and the exporter should be on the label. *Country of origin.* The country where the product originates should appear on the label as well. When the olive oil is treated or packed in another country, the second country may substitute for the first. *Specific origin of olive oil.* In the case of virgin oil, the district of the country from which the olive oil originates is stated on the label. This appears only when the country of origin permits such a distinction. In mixtures of virgin and refined olive oils, only the exporter country should be mentioned. *Batch identity.* The production and packing unit and the batch series number should be noted clearly, on the label. *Date of packing.* The month and the year the product was packed should be shown in numbers. In some countries the month may be written in letters. When the product is packed in December the phrase "End of the year" can be used. *End date of use.* The phrase "Better if used before..." should be on the label. The month and the year should be given in numbers or letters, as mentioned before and the phrase "End of year" could be used instead of December. The period between packing and consumption date should never exceed 12 months. The sooner the oil is used after packing, the better the quality of the product. *Storage by the consumer.* Specific storage conditions, should be mentioned on the container.

Transportation of Large Quantities of Olive Oil

The following information should appear on the containers used for the transportation of olive oil:

- Name of the product
- Net content in weight or volume
- Name and address of the production unit and
- Country of origin from which the product is exported should be included as well.

References

1. Cucurachi, A., Final operations. In Olive oil Technology. Moreno Martinez, J.M., Editor. FAO, Rome (1975).
2. Unal, K., Change in quality of virgin olive oil during storage for 24 months. Ege Univ. Ziraat Fakul. Dergisi. 15:199 (1978).
3. Gutierrez, R.G.Q. and J.M.O. Jimenez, Packaging of olive oil in commercial type containers. II. Packaging in glass, tin polyethylene and polyvinyl chloride containers. Peroxide number evolution. Grasas y Aceites 21:217 (1970).
4. Gutfinger, J., O. Vatariu, M. Alter and A. Leta, Chemical characteristics of Israel olive oils. Grasas y Aceites 26:8 (1975).
5. Kiritsakis, A., Quality studies on olive oil. Ph.D Dissertation. Michigan State University, East Lansing Michigan, USA (1982).
6. Gutierrez, R.G.Q., Bottling and canning. In Olive Oil Technology. Moreno Martinez, J.M., Editor. FAO, Rome (1975).
7. International Olive Oil Council (IOOC). International trade standard applying to olive oils and olive-residue oils. COI/T.15.No 1 (1985).

Chapter 12

Deterioration of Olive Oil

Hydrolytic rancidity or lipolysis and oxidative rancidity or oxidation are the most serious processes leading to deterioration of olive oil. Lipolysis usually commences while the oil is still in the fruit, while oxidation proceeds after the extraction and mainly during storage. The oxidation may occur either in the dark (autoxidation) or in the presence of light (photooxidation).

Hydrolytic Rancidity-Lipolysis

Hydrolytic rancidity is related to the hydrolysis of the glycerides with the consequent increase in total acidity and the deterioration of flavor, caused by certain FFA. Main factors affecting hydrolysis are: moisture, temperature, enzymes and various microorganisms.

The following reaction shows the hydrolysis of a triglyceride, in glycerol and fatty acids. Where R_1,R_2, and R_3 are fatty acid residues.

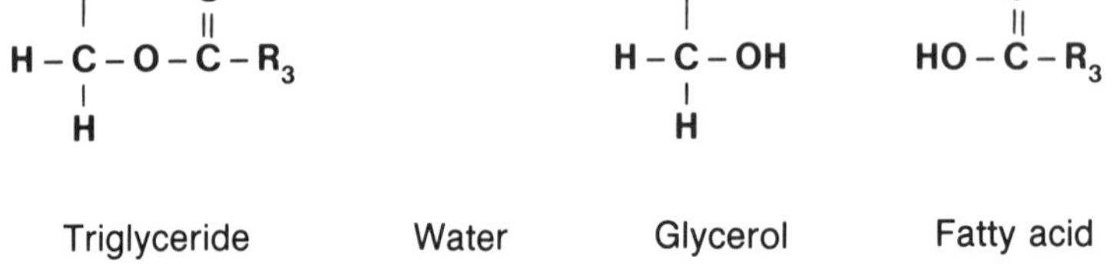

Triglyceride Water Glycerol Fatty acid

Olive oil lipolysis is a result of microbial lipolysis and enzymatic lipolysis.

Microbial Lipolysis

Microbial lipolysis is caused by the microorganisms existing in the olive which liberate the enzyme lipase. From 130 types of bacteria, yeasts and molds isolated, from olives, 70% showed a very high lipolytic activity (1). These microorganisms are so active that they can cause hydrolysis of the glycerides between the time of crushing and the separation of the liquid and solid phases. Improper storage of the fruit (see Chapter 5) favors the formation of microorganisms and the hydrolysis of the olive oil glycerides.

Enzymatic Lipolysis

Enzymatic lipolysis is caused by the natural enzymes (lipases) present in the olive fruit. The endogenous lipase (glyceride-hydrolyzing enzyme) does not manifest its activity until the fruit starts turning purple.

The optimum temperature for lipase action is 45°C and the optimum pH is 8.3. Bruised fruit or fruit damaged by insects displays more lipolytic activity than sound fruit. Fruit remaining on the tree for a long time, or on the ground on the collection nets after natural fall, yield high acidity oil (see more in Chapter 5 and 13). This is due to both the native lipase of the fruit and the lipase of associated microorganisms (2). Table 12.1 shows the change in olive oil acidity during the period fruits remain on the collection nets. If the fruit is stored before processing, and especially if the storage is unsatisfactory (eg piling in thick layers where the fruit is heated by its own respiratory activity), then the combined effect of the endogenous and microbial lipases may result in an increase in the acidity of the oil with a resultant loss of quality. Olive fruit is a live

TABLE 12.1

Increase in olive oil acidity during the period that fruits remain on the collection nets.*

	FFA (%) (expressed as oleic)	
	Olive Fruit Cultivar	
Time (days)	Throumpolia	Tsounati
0	0.6	0.2
10	3.0	-
15	-	0.6
20	4.0	-
30	-	0.8
45	-	2.2
60	-	3.0

*Kiritsakis (3)

breathing organism. Transpiration releases heat which raises the temperature and activates lipolytic enzymes. The presence of water facilitates lipolysis as water dissolves the enzymes and favors microbial growth. This causes undesirable changes in the sensory characteristics of olive oil.

The degree of hydrolytic rancidity is measured by the titration of the liberated fatty acids. The significance of the hydrolytic rancidity is obvious since the quality of olive oil is strongly affected by its acidity (see Chapter 14).

Oxidation

Olive oil is oxidized when it comes in contact with oxygen. Oxidation or oxidative rancidity in olive oil has been known for a long time. It was also known that certain substances retarding oxidation (antioxidants) are present in the plant tissue. Oxidation products have an unpleasant flavor and odor and may adversely affect the nutritional value of the oil. Essential fatty acids, such as linoleic and linolenic, are destroyed and certain fat soluble vitamins disappear. Olive oil is resistant to oxidation (autoxidation) because of the low content of polyunsaturated fatty acids and because it contains natural antioxidants (4,5). However it is very sensitive to photooxidation.

Autoxidation Mechanism

The oxidation mechanism is rather complicated. It is believed that autoxidation proceeds in three stages: initiation, propagation, and termination. According to Dugan (6) the following reactions describe the oil oxidation.

Initiation: In the initiation stage hydrogen is abstracted from an olefinic acid molecule (RH) and a free radical is formed (R·). This radical reacts with an oxygen molecule and a peroxide radical is formed (ROO·). The peroxy radical reacts with another fatty acid molecule (RH) thus producing more peroxides (ROOH) and new free radicals. The

1. **Initiation**

$$RH + O_2 \longrightarrow R\cdot + \cdot OOH$$

Fatty acid → Fatty acid Free radical

2. **Propagation**

$$R\cdot + O_2 \longrightarrow ROO\cdot$$

Peroxy free radical

$$ROO\cdot + RH \longrightarrow R\cdot + ROOH$$

Hydroperoxides

3. **Termination**

$$\left.\begin{array}{l} R\cdot + R\cdot \longrightarrow RR \\ R\cdot + ROO\cdot \longrightarrow ROOR \\ ROO\cdot + ROO\cdot \longrightarrow ROOR + O_2 \end{array}\right] \longrightarrow \text{Nonradical products}$$

RH =	Fatty acid
R·, ROO· =	Free radicals
ROOH =	Hydroperoxides

activation energy required for this process originates either from high temperature storage conditions, or light, or some other source.

At the initiation stage the reactions proceed at a slow rate. A small amount of oxygen is involved. The duration of the initiation stage varies among the different fatty substances (Fig. 12.1) and is related to unsaturation and to the presence or absence of natural antioxidants.

Propagation: It follows the initiation stage at the time when the oil acquires a rancid flavor. The oxidation process is now becoming more complicated since the peroxides formed, being unstable, are easily broken forming more free radicals which in turn participate in new reactions (6). This chain reaction continues until either the unsaturated compound has been exhausted or the free radicals have inactivated each other.

Termination: The mutual annihilation of free radicals is known as the termination stage. The automatic termination of oxidation is difficult, since it is not likely for all free radicals formed to inactivate each other. It is possible however to accelerate the termination, before the oxida-

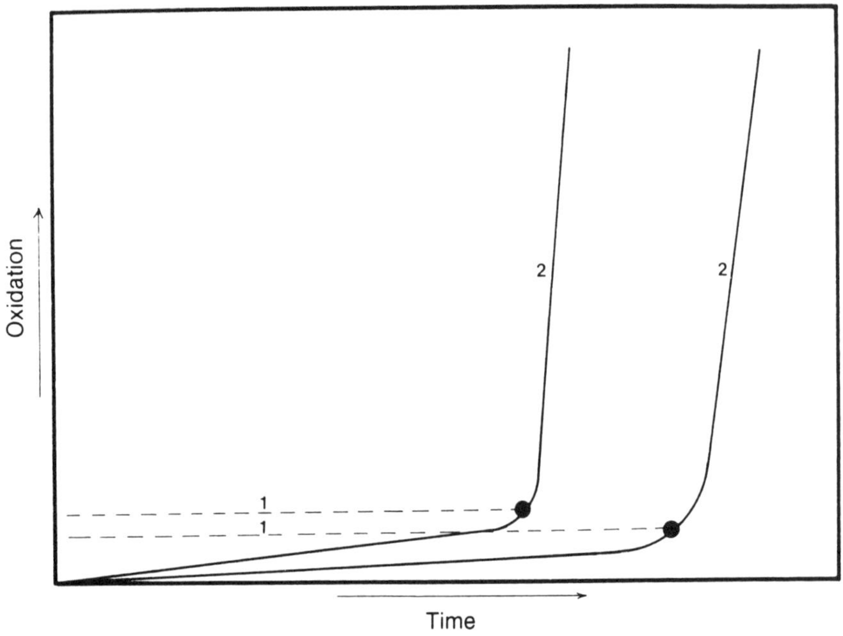

Fig. 12.1. Initiation (1) and Propagation (2) steps in two different olive oils.

tion is advanced, by adding antioxidants, which may inhibit lipid oxidation by reacting with the free radicals and breaking the chain reaction. The antioxidants act as hydrogen donors resulting in the formation of antioxidant free radicals.

AH	+	ROO· ⟶	A·	+	ROOH
Antioxidant		Free radical	Antioxidant free radical		Hydroperoxide

The antioxidant free radicals reacts in two ways.

$$A\cdot + A\cdot \longrightarrow AA$$

$$ROO\cdot + A\cdot \longrightarrow ROOA$$

That is the antioxidant radicals either annihilate each other or react with a peroxide free radical.

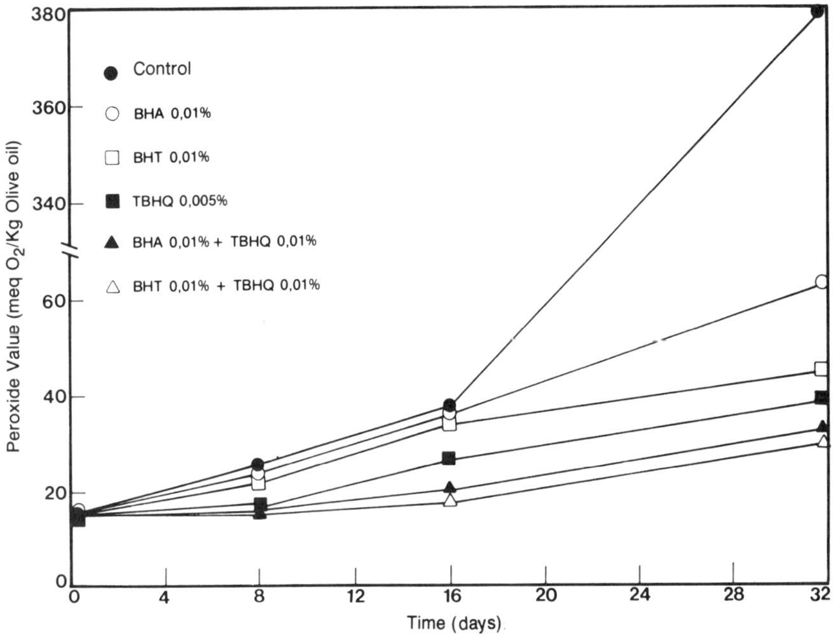

Fig. 12.2. Effectiveness of certain antioxidants in olive oil.

Synthetic antioxidants, such as butylated hydroxyanisole (BHA), butylated hydroxytoluene (BHT), tertiary butylhydroquinone (TBHQ) and propyl gallate gave good results (Fig. 12.2) when added to olive oil (7,8). Kajimoto *et al.* (9) reported that when phospholipids were added to olive oil heated at 180°C, thermal decomposition of the oil tocopherols was avoided. The phospholipids improved the oxidative stability of the heated oil as well. Sheabar and Neeman (10) observed that addition of polyphenols, extracted from olive pomace, to refined olive oil partially inhibited the oxidation of the oil.

Boskou and Katsikas (11) reported that, when the hydrocarbon fraction of olive oil was added to cottonseed oil, the oxidative stability of the latter was increased. The same was true for the triterpenic alcohols which showed a greater protective effect.

Synthetic antioxidants are not permitted to be used in olive oil. Only the natural antioxidant α-tocopherol may be added to refined olive oil, refined olive pomace oil and to olive oil-olive pomace oil mixtures, at the

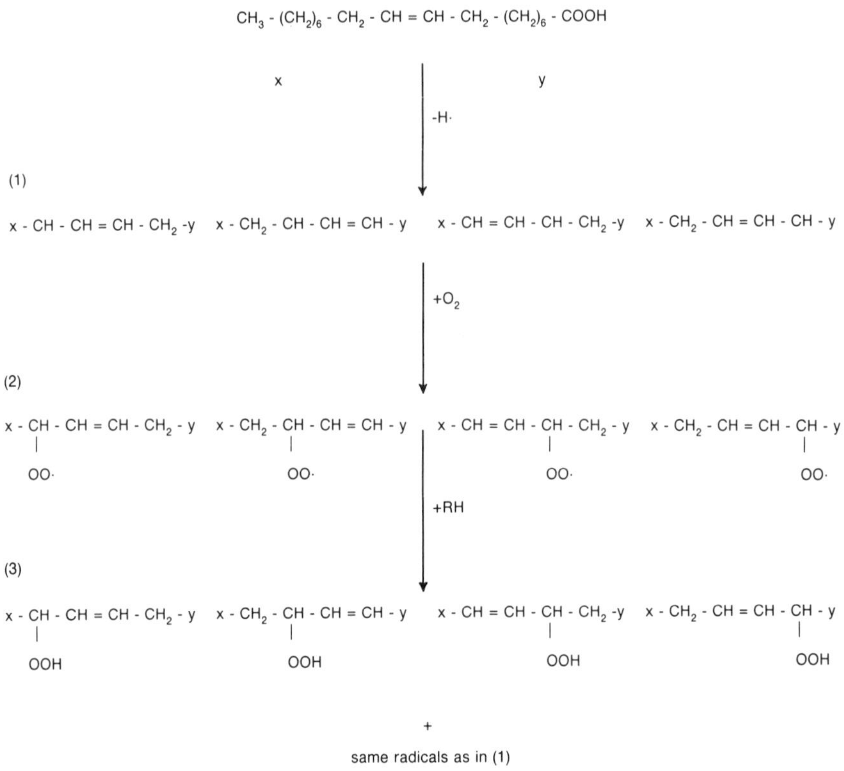

Fig. 12.3. Pathway of oleic acid autoxidation. Note (1) the four possible free radicals (unpaired electron on C-atom 8 or 9 or 10 or 11); (2) the corresponding peroxide radicals; and (3) the related hydroperoxides which form along with new free radicals.

maximum concentration of 200mg/kg. This is to substitute for the natural tocopherol lost during refining.

Oxidation of Oleic Acid

Oxidation of lipids with monoene (oleic) or polyene structures results in a formation of a number of isomeric hydroperoxides (12). Figure 12.3 shows the pathway of oleic acid (present to a high degree in olive oil), autoxidation.

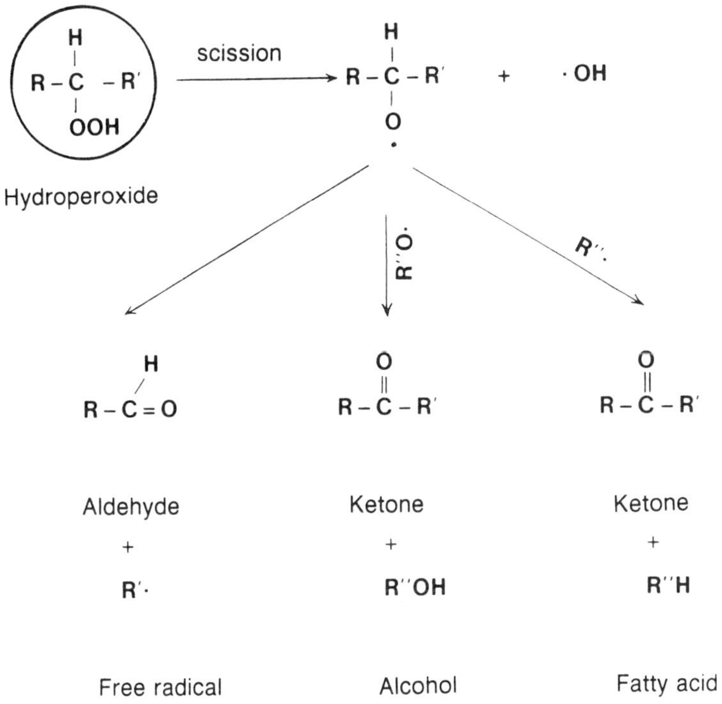

Fig. 12.4. Various compounds formed from hydroperoxide decomposition.

Peroxide Decomposition Products

The peroxides are unstable compounds easily decomposed (6). The decomposition of the peroxides gives a number of volatile compounds (secondary products) with a small number of carbons, responsible for the unpleasant odor and taste, of oxidized oil. Such compounds are aldehydes, ketones and acids of low molecular weight.

Figure 12.4 shows some of the various compounds which may be formed by the decomposition of the primary oxidation products of the olive oil and the other fatty substances. Peroxide decomposition products are the cause of main organoleptic deterioration of the olive oil.

Guitierrez and Romero (13) observed that various compounds are formed during the oxidation of olive oil and it is difficult to differentiate which ones are responsible for the undesirable flavor. It is probable that the aldehydes formed, are mainly responsible for the bad flavor. Yoo *et*

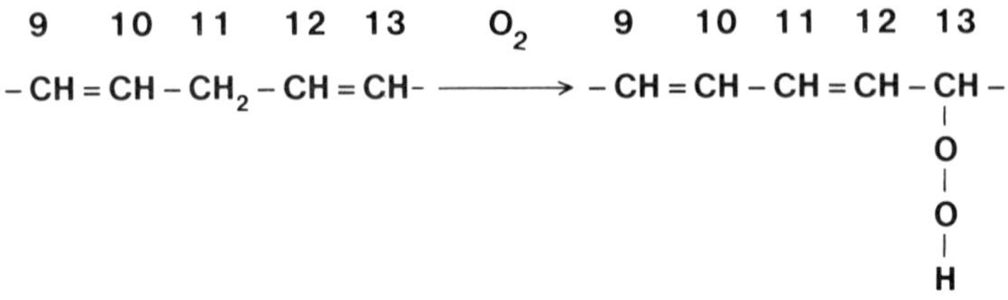

a b

Fig. 12.5. Conjugated and non conjugated double bonds.

al. (14) determined more than 200 volatile compounds produced by heating olive oil in air at 185°C. Nearly 150 of them were identified.

Solinas *et al.* (15) used GLC and HPLC for quantitative determination of the volatile compounds formed during autoxidation of olive oil as determined by organoleptic tests. The volatile fraction was extracted by N_2 stripping. They observed a direct relation between the perceived rancidity and 2-pentenal, hexanal, 2-heptenal, 2-octenal, octanal and nonanal. 2-Pentenal and 2-heptenal were the main indicators of rancidity with thresholds of detection 0.5 and 1.5 ppm respectively.

Formation of Conjugated Hydroperoxides

During the oxidation of linoleic acid (18:2) as well as other polyunsaturated fatty acids displacement of the double bonds and formation of conjugated double bonds takes place (Fig. 12.5). The latter highly absorb in ultraviolet region at 232nm (see Chapter 14). Bartolomeo and Sergio (16) observed that the primary oxidation products of olive oil show an absorption peak at 232nm. For the secondary products (aldehydes, ketones) they found an absorption peak at 270nm. The ultraviolet ratio A232/A270 remains almost constant in virgin olive oils stored in the dark and decreases rapidly in virgin oils exposed to sunlight due to the rapid increase of oxidation (17).

Factors Favoring the Oxidation of Olive Oil

The oxidation of olive oil and other lipids is affected by a number of factors such as: oxygen, temperature, light, ionizing radiation and metals (18).

Oxygen

Oxygen is needed for the oxidation to occur. The oxygen can either be in the "head space" of the container or mixed into the oil. The quantity of oxygen dissolved in the oil varies and depends on the extraction process and the storage and packing conditions. The greater the contact of olive oil with oxygen the more extensive the oxidation. The oxygen dissolved in the olive oil may be 2-2.5% by volume (19). Packing under vacuum or inert gas (eg nitrogen) is very effective in preventing the oxidative rancidity of the olive oil (20,21).

Temperature

Temperature affects the speed of oxidation. High temperature in the storage room accelerates oxidation.

Metals

Certain metallic contaminants, such as copper and iron, have a catalytic effect on the autoxidation of olive oil. Likely sources of these contaminants are the metallic surfaces of storage and processing equipment (see Chapter 10). Reduction and oxidation of the metals occur according to the following reactions:

$$RH + M^{+n} \longrightarrow R\cdot + \cdot H + M^{+(n-1)}$$

$$ROOH + M^{+n} \longrightarrow ROO\cdot + \cdot H + M^{+(n-1)}$$

$$ROOH + M^{+(n-1)} \longrightarrow RO\cdot + \cdot OH + M^{+n}$$

RH =	Fatty acid
ROOH =	Hydroperoxide
R ·, RO ·, ROO · =	free radicals
M^{+n} =	Metal
$M^{+(n-1)}$ =	Metal

The oxidation reduction reaction regulates the catalytic effect of the metals:

$$M^{+2} \longrightarrow M^{+3} + e^-$$

Metals easily oxidized, with the transfer of an electron, are the most active in lipid oxidation.

Vioque (22) proposed a mathematical relationship between olive oil stability and iron concentration of the oil:

$$S = K \times Fe^{\alpha}$$

where, K and α: constants
Fe: iron concentration
S: stability according to the active oxygen method (AOM).

Among all metals found in olive oil, iron appears to be most involved in the oxidation of the oil. The iron originates from the metal surfaces of equipment used for olive oil extraction as well as from the surfaces of storage containers (tanks, drums, etc.). Other metals found in the olive oil are copper and magnesium.

Fedeli *et al.* (23) found that the rate of absorption of oxygen, under accelerated oxidation conditions, was a function of the amount of metal (Ca, Na, Co, Ni, Fe, Cu and Mg) present in the olive oil. Removal of metals by cation exchange, increased the oil resistant to oxidation (22).

FFA

Miyashita and Takagi (24) reported higher oxidation rates for FFA than for their methyl esters. This is attributed to the catalytic effect of the carboxyl groups on the formation of free radicals (25,26).

Olcott (27) reported that addition of oleic acid in refined olive oil, decreased its resistance to oxidation. The effect of antioxidants is considerably decreased in the presence of the FFA (26).

Photooxidation

Photooxidation is a serious deterioration of vegetable oils containing the pigments chlorophylls and pheophytins. These pigments absorb in the visible or near ultraviolet region. Radiations from these regions create electronically excited states.

Chlorophyll and pheophytin, known as sensitizers, exist in two states: *Singlet* and *triplet* states. Oxygen may also exist in a *singlet* or *triplet* state depending on the arrangement of the electrons in the outer orbitals of the molecule.

Mechanism of Photooxidation

Chemical compounds absorb light energy in different wave-length bands depending on the molecular structure. The absorption of radiant energy is known as the primary photochemical process and results in an activated molecule:

$$A + hv \rightarrow A^*$$

Rawls and Van Santen (28) have identified *singlet* oxygen ($^1O_2$) as a source of hydroperoxides. They proposed the following mechanism involving *singlet* oxygen during photooxidation.

$$^1S + hv \rightarrow {}^1S^* \rightarrow {}^3S^*$$

$$^3S^* + {}^3O_2 \rightarrow {}^1O^*_2 + {}^1S$$

$$^1O_2 + RH \rightarrow ROOH$$

$$ROOH \rightarrow \text{Free radical products}$$

(S is sensitizer, superscripts refer to the spin multiplicity and the asterisk indicates electronic excitation).

Singlet oxygen was shown to react directly with methyl linoleate at a rate of 1450 times faster than ground state oxygen (28). They observed that singlet oxygen accounts for approximately 80% of the chlorophyll induced photooxidation. The proton abstraction by the photoactivated carbonyl group of chlorophyll could account for the remaining 20% of the observed photooxidation.

Carlsson *et al.* (29) reported that during exposure of chlorophyll to light the following reaction occurs:

$$Chl \xrightarrow{hv} [Chl]^* \xrightarrow{^3O_2} Chl + {}^1O_2$$

The singlet oxygen then reacts rapidly with C-C unsaturation to give hydroperoxides (30). These hydroperoxides are unstable and decompose into free radicals which initiate the autoxidation chain reaction (29). Although it is generally accepted (28-33) that singlet oxygen is involved during photooxidation, some photosensitized oxidation of lipids may involve the triplet oxygen state (34). Unlike autoxidation, photooxidation with triplet oxygen does not involve chain reactions. Photooxidation does not require an induction period whereas autoxidation involves a long induction period (35).

During autoxidation of polyunsaturated fatty acid via a free radical mechanism, all hydroperoxides formed are conjugated (36). However, when methyl oleate and methyl linoleate are illuminated, in the presence of chlorophyll, both conjugated and nonconjugated hydroperoxides are formed (28,36,37).

Photooxidation of Olive Oil

Taufel *et al.* (38) observed that chlorophyll in the presence of light acts as a prooxidant for methyl oleate. This pigment, however, has no prooxidant effect in the dark. On the contrary, it acts synergistically with phenolic antioxidants.

Interesse *et al.* (39) demonstrated that under the action of light, the four pigments (chlorophylls a and b and pheophytins a and b) present in olive oil develop an oxidizing effect, while in the dark they act as antioxidants. Vazquez *et al.* (40) found that olive oils, because of their chlorophyll content are very sensitive to radiation between 320 and 720nm whether in the presence or absence of antioxidants. Virgin olive oil, due to its chlorophyll content, is very sensitive to light and is easily oxidized (41). Interestingly Valentinis and Romani (42), noted that olive oil exposed to direct sunlight, in the absence of air, had a decrease in the PV during storage. Sanelli (43) studied the quality changes of olive oil in relation to the presence or absence of chlorophyll during storage in closed bottles in darkness or in open bottles in light and in darkness. He found that chlorophyll decreases the formation of FFA and changes the unsaturated/saturated fatty acid ratios during storage. Rahmani (44) reported that the higher the chlorophylls level in olive oil the greater their photocatalytic activity. It should be wise to extract virgin olive oils from ripe fruit, and to remove all leaves prior to processing (44).

Figure 12.6 shows the photooxidation effect of chlorophyll and pheophytin added to bleached olive oil and exposed to fluorescent light.

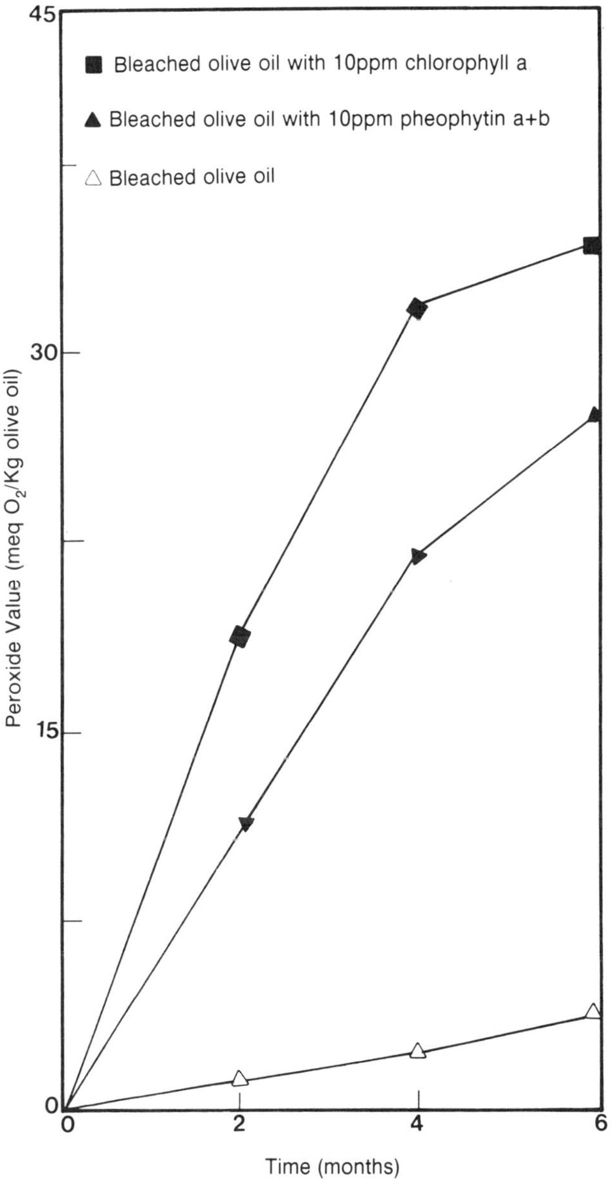

Fig. 12.6. Effect of fluorescent light on photooxidation of bleached olive oil containing added chlorophyll or pheophytin (10ppm).

Kiritsakis *et al.* (7) observed that olive oil exposed to diffused room light was more highly oxidized compared to that protected from light (wrapped with aluminum foil) (Fig. 12.7). Similar observations (Fig. 12.8) were made on olive oil exposed to fluorescent light (45). Figure 12.9 shows that olive oil stored in the dark was oxidized less than that exposed to diffused room light and even less than that exposed to direct sunlight (46). Sastry *et al.* (47) reported that bleaching of chlorophyll occurs via a rather complicated mechanism during photooxidation.

$$\mathrm{Chl} \xrightarrow{h\nu} |\mathrm{Chl}^*| \rightarrow {}^3\mathrm{Chl}$$

$${}^3\mathrm{Chl} + {}^3\mathrm{O_2} \rightarrow {}^1\mathrm{O_2} + \mathrm{Chl}$$

$${}^1\mathrm{O_2} + \mathrm{RH} \rightarrow \mathrm{ROOH}$$

$$\mathrm{ROOH} \xrightarrow{h\nu} \mathrm{ROO\cdot} + \cdot\mathrm{H}$$

$$\mathrm{ROO\cdot} + \mathrm{Chl} \rightarrow \mathrm{ROOH} + \mathrm{Chl\cdot}$$

$$\mathrm{Chl\cdot} + {}^1\mathrm{O_2} \rightarrow \mathrm{Chl\cdot} - \mathrm{O_2}$$

$$\mathrm{Chl\cdot} - \mathrm{O_2} + \cdot\mathrm{H} \rightarrow \mathrm{Chl} - \mathrm{O_2} \quad \text{(Bleached)}$$

The bleaching of chlorophyll is mediated by the hydroperoxide radicals formed during the photooxidation of the oil.

Singlet Oxygen Quenchers

Photosensitized oxidation is not prevented by the antioxidants commonly used to inhibit autoxidation (Fig. 12.10). Photooxidation may be prevented by singlet oxygen quenchers (29,48) which exhibits the opposite effect of sensitizers. The quenchers decrease the rate of oxidation either reacting physically or chemically with the singlet oxygen. The latter is deactivated to the ground state.

Foote and Denny (49) suggested an energy transfer process from singlet oxygen to β-carotene:

$${}^1\mathrm{O_2} + \beta - \mathrm{carotene} \xrightarrow{Kq} \beta - \mathrm{carotene} + {}^3\mathrm{O_2}$$

β-Carotene has been shown to quench singlet oxygen efficiently and thereby inhibit oxidation (29,49,50). Nickel chelates are also known to quench singlet oxygen (45).

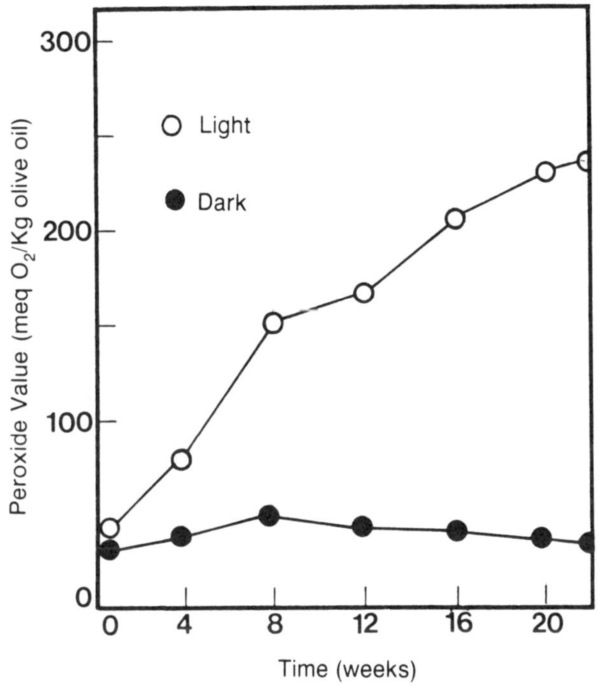

Fig. 12.7 Effect of diffused room light on photooxidation of olive oil.

Carotenoids are well known quenchers of singlet oxygen. Thus photosynthetic organisms are well protected from photooxidation when they contain carotenoids.

Tocopherols function not only as antioxidants in autoxidation, but they are also efficient singlet oxygen scavengers, thereby inhibiting photooxidation. They are scavenging singlet oxygen by either a quenching process or an irreversible reaction. These two pathways are diagrammed as follows (51):

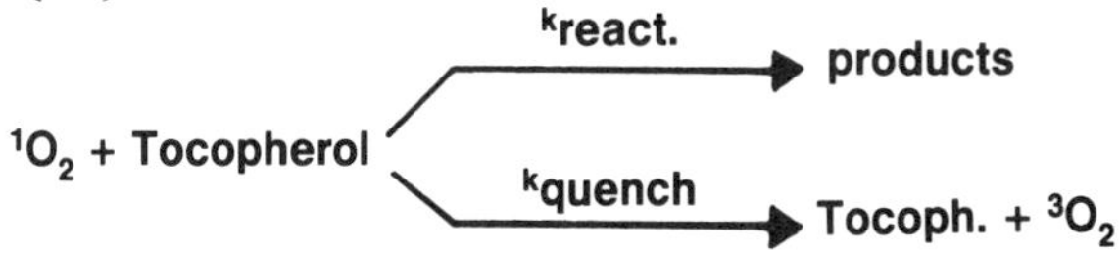

α-Tocopherol is assumed to be the most effective quencher of singlet oxygen among the tocopherols (52) and shows the highest inhibitory

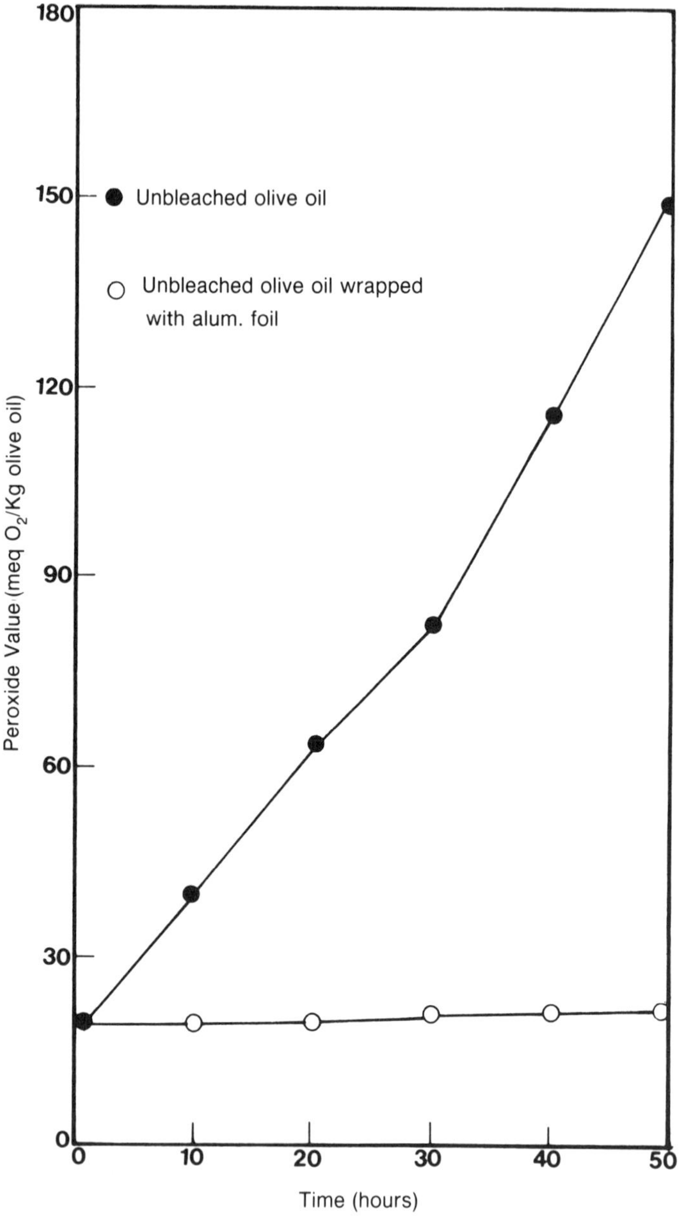

Fig. 12.8. Effect of fluorescent light on the photooxidation of olive oil.

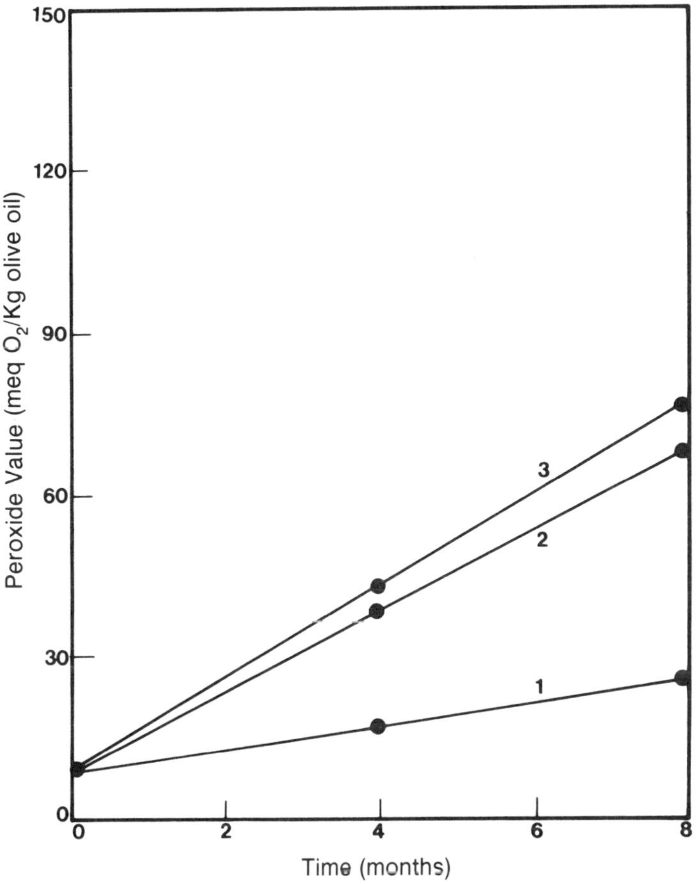

Fig. 12.9. Effect of dark, direct sunlight and diffused room light on the photooxidation of olive oil. [1: Dark; 2: Diffused room light; 3: Direct sunlight (4 hours per day and the rest in diffused light)].

effect in the initial stage of the photooxidation. As the photooxidation proceeds, α-tocopherol is destroyed at a greater rate than γ and δ-tocopherols. Terao *et al.* (53) found that α-tocopherol had weak inhibitory effects on both photooxidation and autoxidation as compared with δ-tocopherols. Combinations of β-carotene and tocopherol exhibited greater protection against photooxidation (54).

Kiritsakis and Dugan (45) studied the effect of α-tocopherol at the concentrations of 50, 100 and 150 ppm, on fluorescent light photooxidized bleached olive oil, at 32°C, and concluded that α-tocopherol had little effect on photooxidation. At a concentration of 100 ppm and in the presence of chlorophyll a (6 ppm), α-tocopherol displayed a prooxidant effect. Tocopherols exhibit a prooxidant effect even during autoxidation (55-57).

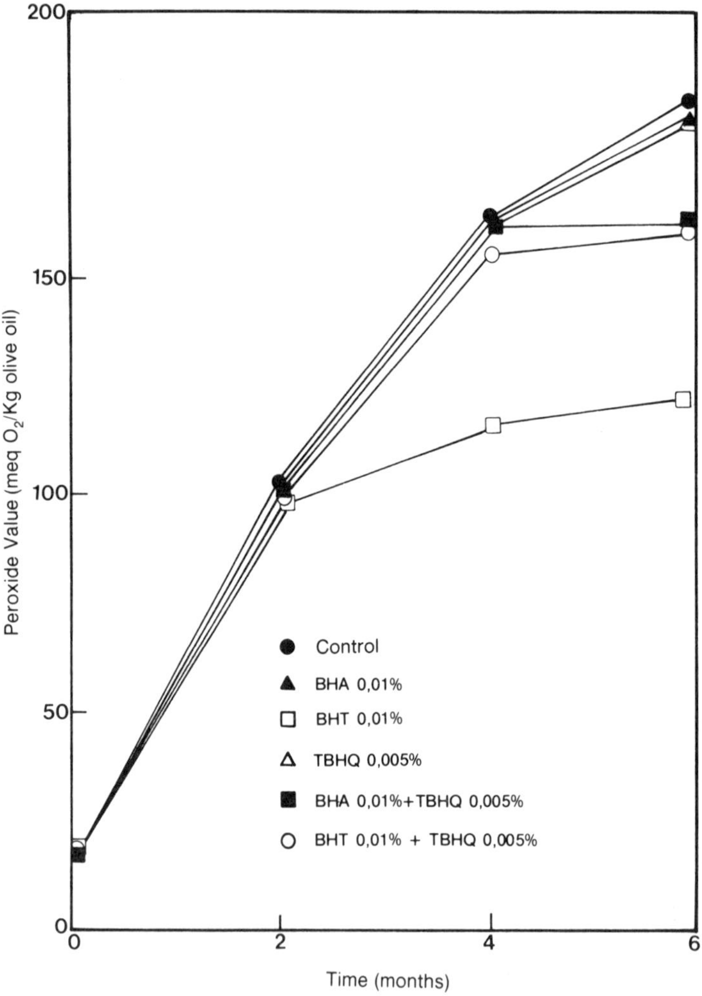

Fig. 12.10. Effectiveness of certain antioxidants against diffused room light.

TABLE 12.2

Peroxide formation in bleached olive oil containing added chlorophyll and other additives during illumination with fluorescent light, at 32°C.*

Treatments	Peroxide Value ($meq\ O_2/kg$ oil)				
	0	8	16	24	32
Bleached (Control)	0	23.0	37.0	48.0	66.0
Bleached + 6ppm chlorophyll	0	27.0	45.0	51.0	72.5
Bleached + 6ppm chlorophyll (wrapped with aluminum foil)	0	4.0	7.0	8.0	8.5
Bleached + 6ppm chlorophyll + 100mg/kg β-carotene	0	5.0	16.0	26.0	57.0
Bleached + 6ppm chlorophyll + 100mg/kg D-α-tocopherol	0	31.0	54.0	63.0	75.0
Bleached + 6ppm chlorophyll + 100mg/kg Ni clelate	0	10.0	16.0	25.0	47.0

*Kiritsakis and Dugan (45)

Table 12.2 shows that olive oil to which carotene and Ni chelates were added was less oxidized than the control, when exposed to fluorescent light.

Other Deteriorations of the Olive Oil

Reversion

Certain oils may develop objectionable flavors. The taste and odor changes occurring are commonly termed "flavor reversion". Principally, oils with linolenic acid components are subject to this change. Although reversion occurs mainly in seed oils (eg soybean, rapeseed) (58), it has been developed in olive oil as well (59). Reversion can be detected even when the PV is much lower (1-4 meq O_2/Kg oil) than that required to produce rancidity (20 meq O_2/kg olive oil). Different factors such as temperature, light, metals and oxygen favor the reversion of flavor. Oxygen is needed for this deterioration to occur.

Clouding of Olive Oil

Clouding of olive oil can be caused by foreign substances which were not removed during clarification or by triglycerides crystallizing out during chilling of the oil. The foreign substances may include water, water-soluble substances and tissue particles. High molecular weight glycerides such as: tripalmitin and tristearin, have relatively high melting points and solidify at refrigeration temperatures. Generally, olive oil extracted with the use of hot water is more readily clarified, than oil extracted with cold water.

Foreign Flavors

Olive oil readily absorbs extraneous flavors. It should therefore be treated and stored in environments free of such flavors. During the extraction, leaves, twigs and other foreign substances which may impart undesirable flavors should be avoided. Fruit damaged by the olive fly (Dacus oleae) also acquires an unpleasant taste (20,60). The presence of iron contributes to an off-flavor in the oil (20). Smoke in the olive oil mill, also contributes to an undesirable flavor and odor in olive oil.

References

1. Suarez Martinez, J.M., Preliminary operations. In Olive Oil Technology. Moreno Martinez, J.M., Editor. FAO, Rome (1975).
2. Kiritsakis, A. and P. Markakis, Effect of olive collection regimen on olive oil quality. J. Sci. Food Agric 35:677 (1984).
3. Kiritsakis, A. 1979. Unpublished work.
4. Notee, E. and N. Romito, Autooxidation of olive oil. Effect of polyphenols. Ind. Agrarie. 9:325 (1971).
5. Vazquez, R.A., J.C. Del Valle, and J. L. M. Del Valle, Polyphenol content and stability of olive oils. Grasas y Aceites. 26:14 (1975).
6. Dugan, L.R., Development and inhibition of oxidative rancidity of foods. Food Technol. 15:10 (1961).
7. Kiritsakis, A., C.M. Stine and L.R. Dugan, The effect of selected antioxidants on the stability of virgin olive oil. M.Sc., Thesis, Michigan State University, U.S.A. (1977).
8. Kiritsakis, A.K., The effect of light on the oxidative deterioration of olive oil. Proceedings of the third International Congress on the Biological Value of Olive Oil. Chanea, Greece (1980).
9. Kajimoto, G., H. Onishi, H. Yoshida and A. Shibahara, Effects of phospholipid on the decomposition of tocopherol in heated vegetable oils. Agric. Chem. Society of Japan. 61:191 (1987).
10. Sheabar, F.Z. and I. Neeman, Separation and concentration of natural antioxidants from the rape of olives. J. Am. Oil Chem. Soc. 65:990 (1988).

11. Boskou, D. and H. Katsikas, Effect of olive oil hydrocarbons and triterpene alcohols on the stability of heated cotton seed oil. Acta Aliment. 8:317 (1979).
12. Dugan, L.R., Lipids. In Principles of Food Science. Part I. Food Chemistry. Fennema, O., Editor. Marcel Dekker Inc., N.Y. (1976).
13. Gutierrez G.Q.R. and A.V. Romero, Estudios sobre el enranciadetermina-cion del grado de rancidez. Grasas y Aceites. 11:67 (1960).
14. Yoo, Y.J., E. Fedeli and W.W. Nawar, The volatile components produced from olive oil by heating. Riv. Ital. delle Sost. Grasse. 65:277 (1988).
15. Solinas, M., F. Angerosa and A. Cucurachi, Relation between the autoxidation products of fats and oils and increase in rancidity by organoleptic evaluation. II. Quantitative determination. Riv. Ital. delle Sost. Grasse. 64:137 (1987).
16. Bartolomeo, D. and R. Sergio, Physicochemical features and acidic composition of some virgin olive oils. Riv. Ital. delle Sost. Grasse. 46:467 (1969).
17. Jimenez, O.J.M. and G.Q.R. Gutierrez, Packaging of olive oil in commercial type containers. III. Preservation in glass, tin, polyethylene, and PVC containers. Changes in absorptivity at 232 and 270nm. Grasas y Aceites. 21:329 (1970).
18. Lea, C.H., The oxidative deterioration of food lipids. In Symposium of Foods, Lipids and their Oxidation. Shultz, H.W., E.A. Day and R.O. Sinnhuber, Editors. AVI Westport, Connecticut (1962).
19. Volpones, E., Solubility of oxygen in olive and seed oils. Quad. Merceol. 13:21 (1972).
20. Cucurachi, A., Final operations. In Olive oil Technology. Moreno Martinez, J.M., Editor. FAO, Rome (1975).
21. Kiritsakis, A. and P. Markakis, The rancidity of olive oil. Geotechnika 4:54 (In Greek) (1978).
22. Vioque, A., Demetalization of edible vegetable oils. A symposium on metal catalyzed lipid oxidation. Marcuse R., Editor. SIK-Rapport, NR 240 (1967).
23. Fedeli, E., A. Brillo and G. Jacini, Metals affecting the autoxidation of vegetable oils. Riv. Ital. delle Sost. Grasse. 50:102 (1973).
24. Miyashita, K. and T. Takagi, Study on the oxidative rate and prooxidant activity of free fatty acids. J. Am. Oil Chem. Soc. 63:1380 (1986).
25. Vega, M.N., J.M.R. Borbolla and Alcala, Studies on olive oil refining. II. The influence of the process on the stability of the oil. Grasas y Aceites. 11:139 (1960).
26. Catalano, M. and M. Felice, Autoxidation of fats. Influence of free fatty acids. Riv. Ital. delle Sost. Grasse. 47:484 (1970).
27. Olcott, S.H., The role of free fatty acids on antioxidant effectiveness in unsaturated oils. J. Am. Oil Chem. Soc. 35:597 (1958).
28. Rawls, H.R. and P.J. Van Santen, A possible role for singlet oxygen in the initiation of fatty acid autoxidation. J. Am. Oil Chem. Soc. 47:121 (1970).
29. Carlsson, D.J., T. Uprunchuk and H.M. Wiles, Photooxidation of unsaturat-

ed oils. Effects of singlet oxygen quenchers. J. Am. Oil Chem. Soc. 53:656 (1976).
30. Kaplan, M.L., "Singlet" oxygen. Chem. Technol. 1:621 (1971).
31. Foote, C.S., Mechanism of photosensitized oxidation. Science. 162:193 (1968).
32. Terao, J. and S. Matsushita, Products formed by photosensitized oxidation of unsaturated fatty acid esters. J. Am. Oil Chem. Soc. 54:234 (1977).
33. Frankel, E.N., W.E. Neff and T.R. Bessler, Analysis of autoxidized fats by gas chromatography mass spectrometry. V. Photosensitized oxidation. Lipids. 14:961 (1979).
34. Chan, H.W.S., Photo-sensitized oxidation of unsaturated fatty acid methyl esters. The identification of different pathways. J. Am. Oil Chem. Soc. 54:100 (1977).
35. Logani, M.K. and R.E. Davies, Lipid oxidation. Biological effects of antioxidants. A review presented at the International Congress on Oilseeds and Oils. New Delhi, India (1979).
36. Khan, N.A., W.E. Tolberg and W.O. Lundberg, Photochemical oxidation of fatty acid esters with and without chlorophyll. Ultraviolet and infra red studies of products. J. Am. Oil Chem. Soc. 31:460 (1954).
37. Terao, J. and S. Matsushita, "The isomeric compositions of monohydroperoxides produced by oxidation of unsaturated fatty acid esters with singlet oxygen". J. Food Proc. Pres. 3:329-337 (1980).
38. Taufel, F., C. Franzke and G. Heber, The role of chlorophyll in oxidation of olefinic fats. Fette seifen Anst. 61:1225 (1959).
39. Interesse, F.S., P. Ruggiero and M. Vitagliano, Autoxidation of olive oil. Effects of chlorophyll pigments. Ind. Agrarie. 9:318 (1971).
40. Vazquez, R.A., J.M.R. Borbolla and Alcala, The influence of light on olive oil alterations. Grasas y Aceites 11:163 (1960).
41. Vitagliano, M., Minor constituents of olive oil. Min. Gras. e Sap. Col. e Vern. 37:136 (1960).
42. Valentinis, G. and B. Romani, Changes of rancidity and peroxide number in edible oil. Boll. Lab. Chim. Prov. 11:351 (1960).
43. Sanelli, B., Keeping quality of vegetable oils as related to their degree of unsaturation and chlorophyll content. Note II. Variations of acidity iodine number, and fatty acid composition. Riv. Ital. delle Sost. Grasse. 58:184 (1981).
44. Rahmani, M., Review of the role of chlorophyll pigments in the photooxidation of virgin olive oil. Olivae. 6:30 (1989).
45. Kiritsakis, A.K. and L.R. Dugan, Studies in photooxidation of olive oil. J. Am. Oil Chem. Soc. 62:892 (1985).
46. Kiritsakis, A., Koutsaftakis, A. and L. Miktos, The effect of processing systems PIERALISI, HILLER AND RAPANELLI (Sinolea-Decanter) on olive oil quality. Grasas y Aceites. 36:165 (1985).
47. Sastry, Y.S.R., P.V. Rao and G. Lakshminarayana, Bleaching behavior of

chlorophyll substrates under vacuum on exposure to visible light. Oleagineus. 28:467 (1963).
48. Foote, C.S., V.C. Chang and R.W. Denny, Chemistry of singlet oxygen. X. Carotenoid quenching parallels biological protection. J. Am. Oil Chem. Soc. 92:5216 (1970).
49. Foote, C.S. and R.W. Denny, Chemistry of singlet oxygen. VII. Quenching by β-carotene. J. Am. Oil Chem. Soc. 90:6233 (1968).
50. Matsushita, S. and J. Terao, Singlet oxygen initiated photooxidation of unsaturated fatty acid esters and inhibitory effects of tocopherols and β-carotene. In Autoxidation in foods and biological systems. Michael G. Simic and Marcus Karel, Editors. Plenum Press, New York and London (1980).
51. Houlihan, C. and Chi Tang Ho, Natural Antioxidants. In Flavor chemistry of fats and oils. Min, O. and T. Smouse, Editors. (AOCS), Champaign, Ill (1985).
52. Yamauchi, R. and S. Matsushita, "Quenching effect of tocopherols on the metyl linoleate photooxidation and their oxidation products", Agri. Biol. Chem. 41:1425 (1977).
53. Terao, T., R. Yamauchi, H. Murakani and S. Matsushita, Inhibitory effects of tocopherols and β-carotene on singlet oxygen initiated photooxidation of methyl linoleate and soybean oil. J. Food Proc. Pres. 4:79 (1980).
54. Min, D.B. and E.C. Lee, Factors affecting the singlet oxygen oxidation of soybean oil. Fifth International Flavor Conference. Frontiers of flavor. Halkidiki, Greece (1987).
55. Cillard, J. and P. Cillard, Inhibitors of the prooxidant activity of α-Tocopherol. J. Am. Oil Chem. Soc. 63:1165 (1986).
56. Cillard, J., P. Cillard and M. Cormier, Effect of experimental factors on the prooxidant behavior of α-Tocopherol. J. Am. Oil Chem. Soc. 57:255 (1980).
57. Husain, R., J. Cillard and P. Cillard, α-Tocopherol prooxidant effect and malonaldehyde production. J. Am. Oil Chem. Soc. 64:109 (1987).
58. Lundberg, W.O., Autoxidation and Antioxidants. Lundberg. W.O., Editor. Interscience Publishers, Wiley & Sons, Inc., N.Y. (1962).
59. Gutierrez, G.Q.R., Ensayos sobre el empleo de "antioxidantes" en la conservacion de oliva "fino" y envasado en latas. Grasas y Aceites. 14:50 (1963).
60. Newenschuander, P. and S. Michelakis, The infestation of dacus oleae at harvesting time and its influence on yield and quality of olive oil in Crete. Zet. Ang. Entom. 86:420 (1978).

Chapter 13

Factors Affecting Olive Oil Quality

Many factors affect the quality of olive oil. These factors can be categorized as those which act during the formation of the oil in fruit, factors acting from the formation of the oil in fruit to collection of the fruit, factors acting during the storage of the fruit, factors acting during the extraction process, and factors acting during the storage of olive oil.

Factors Acting During the Formation of the Oil in Fruit

Among the most important factors of this group are the olive cultivar, the soil and the climate, where the trees grow. Certain cultivars produce oil of more desirable sensory qualities than others. Drained and calciferous soils rather than wet and clay soils also yield oil of better sensory characteristics (1). Dry climates with a lot of sunshine produce oil of better flavor (see Chapter 3).

Factors Acting from the Formation of the Olive Oil in the Fruit to Collection of the Fruit

Insects, microorganisms, or other environmental agents, may affect the quality of the oil during the period from its formation on the fruit to collection of the fruit (2-5). Lipolytic enzymes are involved in many undesirable changes (6,7).

The acidity of olive oil is increasing progressively even when the fruit remain on the tree due to the activation of the enzyme lipase. The acidity of the oil is also increased during the period the fruit remain on the collection nets (see Chapter 5). The reason for the development of acidity in the olives left long on the nets (Fig. 13.1), is that both the endogenous lipases of the fruit and the lipases of microorganisms that may grow on the fruit are likely to effect the lipolysis. Bruising sustained by the olives as a result of the fall facilitates the contact of the fruit lipases with their substrates; and the conditions in the fruit pile (rising temperature and humidity, broken skin) are conducive to the growth of microorganisms.

Table 13.1 summarizes the data relating olive collection procedures and the PV of olive oil. The values indicate a more rapid deterioration for

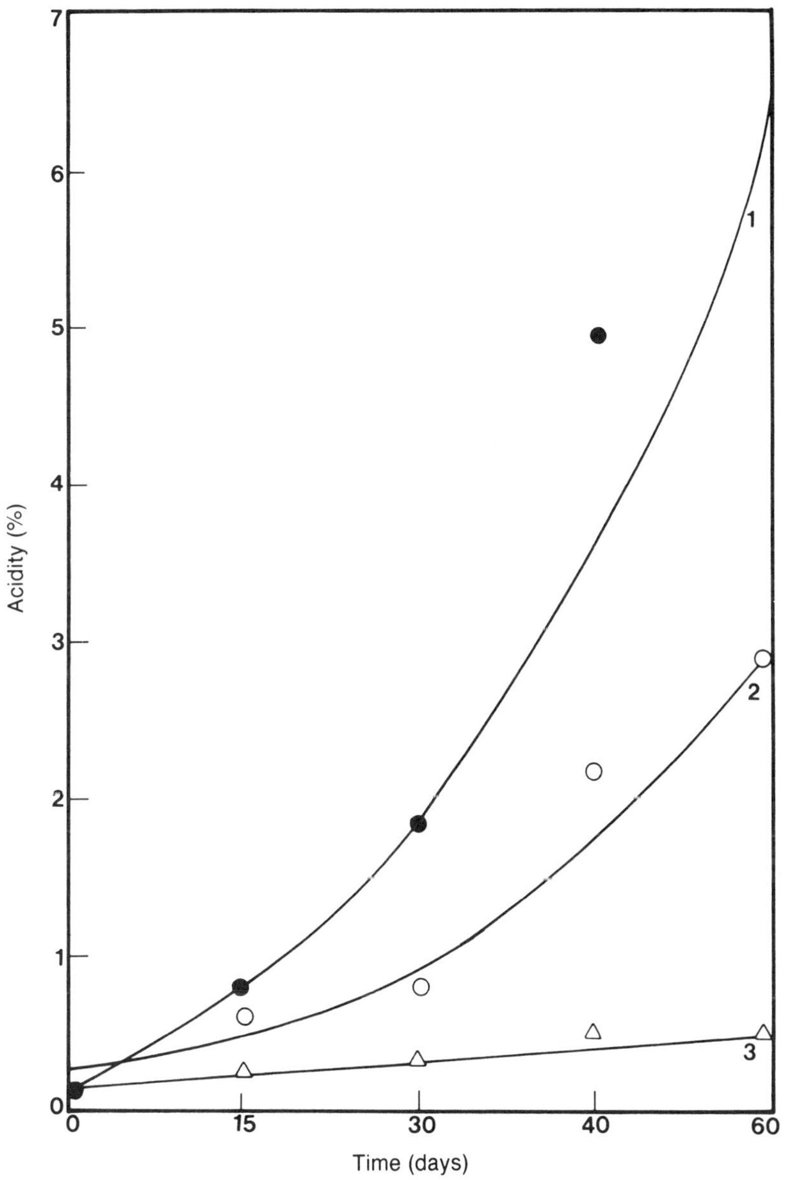

Fig. 13.1. Free fatty acid content (% oleic acid) of olive oil as affected by the time and method of fruit collection. (△: Olive fruit collected, from plastic nets after shaking the trees. ○: Olive fruit collected, from nets after natural falling. ●: Olive fruit collected, direct from the tree).

TABLE 13.1

Peroxide value (meq O_2/Kg olive oil) as affected by time and the method of fruit collection.*

	Collection (days)			
Origin of fruit sample	0	15	45	60
Tree	5.7	5.7	10.0	7.0
Upper net layer (natural fall)	-	10.8	32.7	22.5
Lower net layer (after shaking)	6.0	11.6	47.7	43.8

*Kiritsakis and Markakis (8)

the fruit resting on the nets than for the fruit remaining in the tree. The PV apparently reaches a maximum at about 6 weeks after shaking (lower net layer) and then decreases, probably as the aliphatic hydroperoxides commence degrading (see Chapter 12).

Both autoxidation and enzymic lipid oxidation are likely to occur in the fruit during the time from oil formation to fruit collection.

Factors Acting During the Storage of the Olive Fruit

The effect of storage conditions on oil quality has already been discussed in details (see Chapter 5).

Factors Acting During Extraction of Olive Oil

It has already been mentioned that the conditions of extraction and processing affect the quality of the oil (see Chapters 6 and 7).

Factors Acting During Storage of the Olive Oil

The deterioration of olive oil quality, during storage, is favored by factors such as: oxygen, temperature, metals, light and others (see Chapter 12). In order to obtain the best possible quality of olive oil (low acidity, negligible oxidation and best sensory characteristics) the following suggestions appear in order:

- Protect the fruit while it is on the tree. Collect it with a minimum of damage and store it for the shortest possible time and under favorable conditions (cool places) before processing. (Olives harvested by nets subtended to the trees must be subjected to oil extraction as quickly as possible and not longer than 2 or 3 weeks).

- Use stainless steel equipment for processing avoiding high temperature and excessive exposure to air.
- Store and pack the oil in proper containers, at low temperatures, under a minimum contact with air and exposure to light (packing under vacuum or inert gas) and
- Possible enrichment of olive oil with natural antioxidants, such as α-tocopherol and phenols (tyrosol and hydroxytyrosol).

References

1. Papanastassiou, D.P., Modern Olive Oil Processing. Olive oil technology. Spiros-Spirou and Son, Editors. Athens, Greece (in Greek) (1966).
2. Cucurachi, A., Final operations. In Olive Oil Technology. Moreno Martinez, J.M., Editor. FAO, Rome (1975).
3. Newenschuander, P. and S. Michelakis, The infestation of *dacus oleae* at harvesting time and its influence on yield and quality of olive oil in Crete. Zet. Ang. Entom. 86:420 (1978).
4. Sagasta, M.E., Maux et maladies de l' olivier. In Manuel d' oleiculture. Humanes, J.G. and J.M. Phillipe, Editors. FAO, Rome (1976).
5. Kapatos, E.T. and B.S. Fletcher, An assessment of components of crop loss due to infestation by *Dacus Oleae* in Corfu. Entom. Hellenica. 1:7 (1983).
6. Kiritsakis, A. and P. Markakis, The rancidity of olive oil. Geotechnika. 4:54 (in Greek) (1978).
7. Kiritsakis, A. and P. Markakis, Olive oil. A review. Advances in Food Research. 31:453 (1987).
8. Kiritsakis, A. and P. Markakis, Effect of olive collection regimen on olive oil quality. J. Sci. Food Agric. 35:677 (1984).

Chapter 14

International Olive Oil Council-Olive Oil Quality

The international olive oil council (IOOC) was created by parties interested in olive oil as an international trade commodity. Currently, the council has many member nations accounting for 96% of the world's olive oil production.

Olive Oil Agreement

The first international olive oil agreement was proposed in 1956 and it was amended in 1958. Other agreements, in 1963, 1979, and 1986, followed. The general objectives of the last agreement (1) on olive oil and table olives are:

1. With respect to international co-operation and concerted action:
   - To foster international co-operation for the integrated development of the world economy for oil products;
   - To maintain fair working conditions throughout the oil-growing and oil-products industry in order to raise the standard of living of the populations engaged therein;
   - To foster the co-ordination of production, industrialization and marketing policies for olive oil, oil- pomace oils and table olives;
   - To study and facilitate the application of the necessary measure for the other products of the oil tree;
   - To continue and extend the work accomplished under the previous international olive oil agreements.
2. With respect to the modernization of oil cultivation and olive-oil extraction:
   - To encourage research and development in order to elaborate techniques that could modernize oil husbandry and the olive-products, industry through technical and scientific planning; improve the quality of olive products, reduce the production cost of olive oil, improve the situation of the oil-products industry in accordance with the recommendations of the United Nations Conference on the Human Environment, to abate any harmful effects;
3. With respect to the expansion of international trade in oil products:
   - To facilitate the study and application of measures designed to expand international trade in oil products.

- To adopt whatever measures are appropriate to develop the consumption of olive oil and table olives;
- To forestall and, where appropriate, combat any practices of unfair competition in international trade in olive oil, oil- pomace oils and table olives.
- To improve market access and reliability of supply;
- To undertake any activities and measures that could highlight the significant effect of olive oil and table olives on health.

4. With respect to the standardization of international trade in olive products:
   - To facilitate the study and application of measures for balancing production and consumption;
   - To facilitate the study and application of measures for harmonizing national laws relating, in particular, to the marketing of olive oil and table olives;
   - To lessen the drawbacks associated with fluctuations in the availability of supplies on the market, in order to prevent excessive fluctuations in prices, and to create conditions that will enable productions, consumption and international trade to expand harmoniously;
   - To improve procedures for information and consultation in order, to enhance the market for olive oil, oil- pomace oils and table olives.

The IOOC has organized three international congresses on the nutritional value of olive oil and its effect on health (see Chapter 17).

Classification of Olive Oil

At a 1985 meeting the IOOC proposed the following designations for olive oil and oil pomace oil.

1. **Virgin olive oil:** is the oil obtained from the fruit of the oil tree (Olea europaea sativa Hoff.Link) only by mechanical or other physical means under conditions, particularly thermal, that do not lead to alterations in the oil, and which has not undergone treatment other than washing, decantation, centrifugation, and filtration.

Virgin olive oil includes the following: *Extra virgin olive oil:* Virgin olive oil of absolutely perfect flavor and odor having a maximum acidity, in terms of oleic acid, of 1 g/100g. *Fine virgin olive oil:* Virgin olive oil of absolutely perfect flavor and odor having a maximum acidity, in terms of oleic acid, of 2.0g/100g. *Semi-fine virgin olive oil (or ordinary virgin olive oil):* Virgin olive oil of good flavor and odor having a maximum acidity, in terms of oleic acid, of 3.3g/100g with a margin of tolerance 10%. *Virgin olive oil not fit for consumption as it is:* designated virgin olive oil lampante. This oil should undergo refining or will be used for technical purposes. Is an off flavor and/or off-smelling virgin olive oil with a maximum acidity, in terms of oleic acid, of more than 33g/100g.

2. **Refined olive oil:** is the oil obtained from virgin olive oils by refining methods which do not lead to alterations in the initial glyceride structure, having a maximum acidity in terms of oleic acid, of 0.5g/100g.

3. **Olive oil:** is the oil consisting of a blend of virgin olive oil (except lampante) and refined olive oil, having a maximum acidity in terms of oleic acid, of 1.5g/100g.

4. **Crude olive-pomace oil:** is the oil obtained by treating olive pomace with solvents, to the exclusion of oils obtained by reesterification processes and any mixture of other oils.

5. **Refined oil- pomace oil:** is that obtained from crude olive-pomace oil by a refining process, having a maximum acidity in terms of oleic acid of 0.5g/100g.

6. **Olive pomace oil:** is a mixture of refined olive-pomace oil and virgin olive oil (except lampante), having a maximum acidity in terms of oleic acid, of 1.5g/100g.

Table 14.1 shows the criteria set by the IOOC for the differentiation of olive oils, besides the definitions given above:

TABLE 14.1

Criteria for the differentiation of olive oils.

	Result
1. Semidrying Oil Test	
Virgin olive oil)	
Refined olive oil)	negative
Refined olive-pomace oil)	
2. Olive-pomace Oil Test	
Virgin olive oil)	
Refined olive oil)	negative
Refined olive-pomace oil)	not relevant
3. Cottonseed Oil Test	
Virgin olive oil)	
Refined olive oil)	negative
Refined olive-pomace oil)	

TABLE 14.1 continued

Criteria for the differentiation of olive oils.

	Result
4. Teaseed Oil Test	
Virgin olive oil)	
Refined olive oil)	negative
Refined olive-pomace oil)	
5. Sesameseed Oil Tests	
Virgin olive oil)	
Refined olive oil)	negative
Refined olive-pomacc oil)	
6. Relative Density (20°C/water at 20°C)	
Virgin olive oil)	
Refined olive oil)	0.910-0.916
Refined olive-pomace oil)	
7. Refractive Index (n_D20°C)	
Virgin olive oil)	
Refined olive oil)	1.4677-1.4705
Refined olive-pomace oil)	1.4680-1.4704
8. Saponification Value (mg KOH/g oil)	
Virgin olive oil)	
Refined olive oil)	184-196
Refined olive-pomace oil)	182-193
9. Iodine Value (Wijs method)	
Virgin olive oil)	
Refined olive oil)	75-94
Refined olive-pomace oil)	75-92
10. Unsaponifiable Matter (using light petroleum)	
Virgin olive oil)	
Refined olive oil)	not more than 15 g/kg
Refined olive-pomace oil)	not more than 25 g/kg
11. Bellier Index*	
Virgin olive oil)	
Refined olive oil)	not more than 17
Refined olive-pomace oil)	not applicable

*If the value is higher than 17, the oleic, arachidic, vehenic and lignoceric content, must be reported.

The relationships between the aforementioned designations and certain sensory and laboratory criteria are shown in Tables 14.2 and 14.3.

TABLE 14.2

Designations of Olive Oil as Related to Sensory and Laboratory Criteria.*

	Olive Oil					
Sensory Characteristics	Virgin olive oil extra	Virgin olive oil fine	Virgin olive oil semi-fine	Virgin olive oil lampante	Refined Olive oil	Olive Oil
Odor	Absolutely perfect	Absolutely perfect	Good	Off smelling	Acceptable	Good
Taste	Absolutely perfect	Absolutely perfect	Good	Off smelling	Acceptable	Good
Color	Light yellow to green	Light yellow to green	Light yellow to green	Yellow to brownish green	Light yellow	Light yellow to green
Appearance at 20°C	Limpid	Limpid	Limpid	—	Limpid	Limpid
FFA in oleic acid (%) (max)	≤1.0	≤2.0	≤3.3	>0.5	≤0.5	≤1.5
PV (max) (meq O_2/kg oil)	≤20	≤20	≤20	>20	≤10	≤20
Moisture and volatile matter (%) (max)	≤0.2	≤0.2	≤0.2	≤0.3	≤0.1	≤0.1
Impurit. insol. in light petrol. (%) (max)	≤0.1	≤0.1	≤0.1	≤0.2	≤0.05	≤0.05
K (max) 270	≤0.20	≤0.20	≤0.30	≤0.30	≤1.10	≤0.90
ΔK (max)	≤0.01	≤0.01	≤0.01	≤0.01	≤0.15	≤0.15

*IOOC (2,3)

K_{270} is absorbance $E^{1\%}_{1\,cm}$ at 270 nm

TABLE 14.3

Designations of Olive Pomace Oil as Related to Sensory and Laboratory Criteria.*

Sensory Characteristics	Olive	Pomace	Oil
	Crude olive pomace oil	Refined olive pomace oil	Refined olive pomace oil and olive oil
Odor	—	Acceptable	Acceptable
Taste	—	Acceptable	Aceptable
Color	—	Light yellow to brownish yellow	Light yellow to green
Appearance at 20°C	—	Limpid	Limpid
FFA in oleic acid (%) (max)	No limit	≤0.5	≤1.5
PV (max) (meq O_2/kg oil)	No limit	≤20	≤20
Moisture and volatile matter (%) (max)	≤1.5	≤0.1	≤0.1
Impurities insol. in light petrol. (%) (max)	—	≤0.05	≤0.05
Soap test	—	Negative	—
K (max) 270	—	≤2.00	≤1.70
ΔK (max)	—	≤0.20	≤0.18

*IOOC (2,3)

Quality Criteria of Olive Oil

The acidity, the oxidation, the color and the sensory characteristics (odor and taste) are the basic criteria for grading olive oil. According to Cucurachi (4), grading based on the sensory characteristics is the most important one.

Acidity

Acidity is the most basic criterion of grading olive oil. Total acidity is essentially a measure of the FFA present in the oil. It is categorized into edible (acidity ≤ 3.3%) and industrial (acidity more the 3.3%). An apparatus called FIAtron (5) is commercially available for the rapid determination of total acidity. It uses n-butanol/water as a solvent and methyl red as indicator. The agreement between FIAtron procedure and the conventional titration is good.

Oxidation

The measurement of the degree of oxidation is another criterion for evaluating the quality of the olive oil. Oxidation is determined by measuring the PV, the ultraviolet (UV) absorbance or other parameters.

Peroxide Value

The most commonly used method for measuring the total hydroperoxides formed as the primary oxidation products is the PV procedure (6).

Maurikos *et al.* (7) used a polarographic method for determining the PV in virgin olive oil. The supporting electrolyte was LiCL in methanol benzene with a dropping mercury electrode. For virgin olive oil the peroxide value must be <20 (see Table 14.2).

Ultra Violet Absorbance

UV absorbance measurements are an index for the degree of oxidation and adulteration of olive oil (see Chapter 15). Many workers (8-11) reported that there is a good correlation between the absorbance of olive oil at 232nm and to the degree of oxidation.

Specific Coefficients of olive oil

The values of the specific coefficients

$(K^{1\%}_{1cm}232, K^{1\%}_{1cm}270, \text{and} \Delta K)$ of olive oil are used for evaluating its quality.

The conjugated hydroperoxides absorb at 232nm while at other wave lengths (262,268,270, and 274nm) the secondary oxidation products (aldehydes and ketones) absorb. Conjugated diene and trienes, formed during refining or bleaching of olive oil, also absorb at 270nm. Thus high absorbance at 270nm is related to olive oil oxidation or to refining process or to both of them. Low absorbance values (K_{232}, K_{270}, and ΔK) correspond to good olive oil quality.

The IOOC has proposed special limits for these coefficients for every category of olive oil. The value:

$$\Delta K = K_{268} - \frac{K_{262} + K_{274}}{2}$$

is used to differentiate extra, fine and semi-fine olive oils (see Table 14.2).

Olive oil with a specific absorbance (K270)> 0.20is considered a virgin, if their absorbance after passing through the active alumina is <0.11.

Methods Measuring the Secondary Oxidation Products of Olive Oil

For detecting the secondary oxidation products of the olive oil, Kreis test and TBA test, may be applied. The Kreis test has been applied in commercial practice for a long time. This procedure depends upon a reaction between phloroglucinol and secondary oxidation products of the oil such as epihydrin aldehyde and other homologues of this compound.

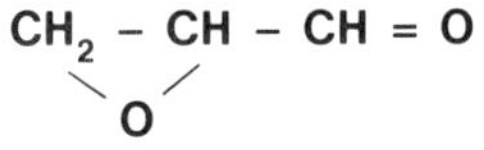

The TBA test has been applied particularly in cooked meats, although it is frequently used to estimate the degree of oxidation of other lipid foods (12). TBA test has been used by several workers (13-15) to measure the oxidative rancidity of olive oil. Casillo (14) reported that the TBA test detects the rancidity of olive oil at a lower level than other tests (PV, Kreis test). Malonaldehyde forms readily when trienoic or tetraenoic acids are present (12). It may also be formed from dienoic acid (linoleic) through the following stages:

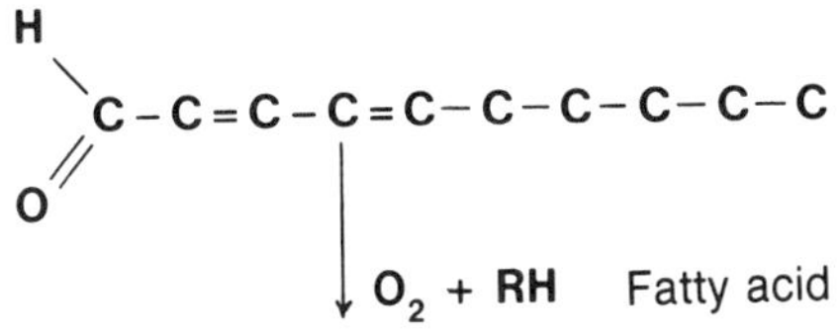

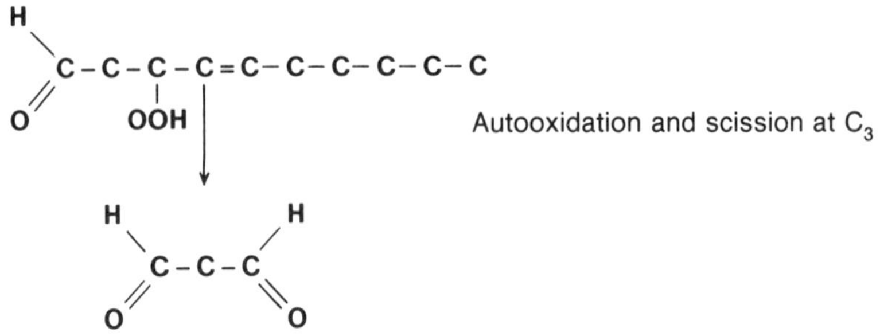

Malonaldehyde

A colorimetric reaction occurs during the application of TBA test and a chromagen is formed through the condensation of two molecules of TBA with one molecule of malonaldehyde.

The determination of the oxidative rancidity of olive oil should not be based only in one method, since each one of them has a weak point. The use of two or more methods gives a better indication for the oxidative deterioration of the oil.

Color

The intensity of the color formed is used as an indication of the degree of oxidation of olive oil. The color of olive oil is a quality indicator. It may vary among different olive oils extracted at different times. The pigments (chlorophylls, pheophytins, xanthophylls, and carotenes) mostly present in the olive fruit, at harvesting time, are responsible for the color of the olive oil. It is green at the beginning of the season when the olive fruit is still immature and the presence of chlorophyll is dominant. As maturity advances the oil turns golden yellow because of the carotenes present, while overripe fruit gives olive oil of a green-light brown color, mainly due to pheophytins. The method of oil extraction also affects the oil color. The Rapanelli-decanter oil is greener than the Rapanelli-sinolea oil because the latter contains less chlorophyll (15-18).

Sensory Evaluation of Virgin Olive Oil

Sensory characteristics, of the oil is another criterion for evaluating its quality. Based on flavor the following classification of olive oil was proposed by Frezzotti and Manni (19).

- Unripe oil-extracted from unripe olives.
- Bitter oil-extracted from olives mixed with leaves.
- Fruity oil-extracted from freshly picked, ripe oils (best olive oil).
- Good flavor oil-ordinary oil with off flavors.
- Defective oil- olive oil with off-flavors (mouldy, wormy, rancid, metallic, medicinal, etc.).

The IOOC (20) proposed the following vocabulary, in alphabetical order, for taste and flavor of virgin olive oil used for sensory analysis.

Almond: This flavor may appear in two forms; that typical of the fresh almond or that peculiar to dried, sound almonds which can be confused with incipient rancidity. A distinctive taste is perceived when the oil remains in contact with the tongue and the palate.

Apple: Flavor of olive oil which is reminiscent of this fruit.

Atrojado (fusty): Characteristic flavor of oil obtained from oils stored in piles which have undergone an advanced stage of fermentation.

Bitter: Characteristic taste of oils obtained from green oil fruit or olives turning color. It can be more or less pleasant depending on its intensity.

Brine: Flavor of oil extracted from oil fruit preserve in saline solutions.

Cucumber: Flavor produced when an oil is hermetically packed for too long, particularly in tin containers. This is attributed to the formation of the compound 2,6 nonadienal.

Earthy: Characteristic flavor of oil obtained from oil fruit which have been collected with earth or mud on it which has not been washed away. This flavor may sometimes be accompanied by a musty-humid flavor.

Esparto: Characteristic flavor of oil obtained from olives pressed in new esparto mats. The flavor may differ depending on whether the mats are made of green esparto or dried esparto.

Flat or smooth: Flavor of olive oil with very weak sensory characteristics due to the loss of their aromatic compounds.

Fruity: Flavor which is reminiscent of both the odor and taste of sound, fresh fruit picked at its optimum stage of ripeness.

Grass: Characteristic flavor of certain olive oils reminiscent of recently mown grass.

Green leaves (bitter): Flavor of oil obtained from excessively green oils or oils that have been crushed with leaves and twigs.

Grubby: Characteristic flavor of oil obtained from oils which have been heavily attacked by the oil fly (Dacus oleae).

Harsh: Characteristic sensation of certain oils reminiscent of more or less dried grass.

Heated or burnt: Characteristic flavor of oils caused by excessive and/

or prolonged heating during processing, particularly when the paste is thermally processed.

Metallic: Flavor that is reminiscent of metal. Characteristic of oils which have been in prolonged contact, under unsuitable conditions, with foodstuffs or metallic surfaces during crushing, mixing, pressing or storage.

Muddy sediment: Characteristic flavor of oil recovered from the decanted sediment in vats and underground tanks.

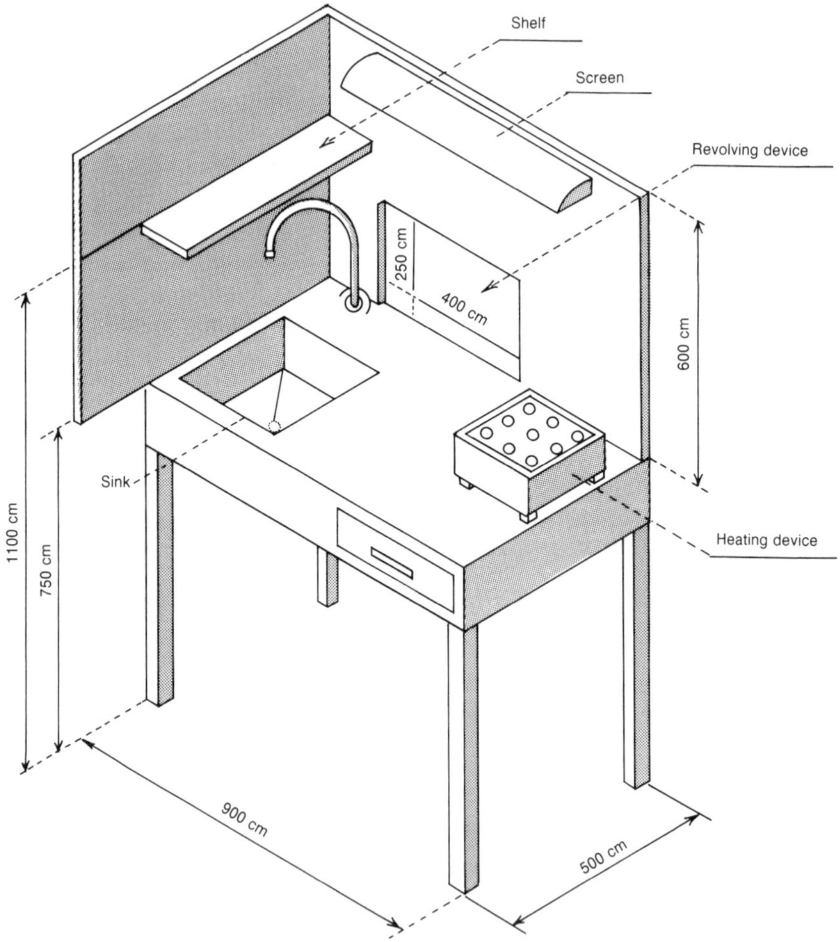

Fig. 14.1 Arrangement of booth for olive oil sensory analysis.

Mustiness-Humidity: Characteristic flavor of oils obtained from fruit in which large numbers of fungi and yeasts have been developed. This is mainly due to storage in piles, in humid conditions, for several days.
Old: Characteristic flavor of oil that has been kept too long in storage containers. This flavor also appears in oils which have been packed for an excessively long period.
Pomace: Characteristic flavor that is reminiscent of the flavor of oil pomace.
Pressing mat: Characteristic flavor of oil obtained from olives that have been pressed in dirty mats, where fermented residues have been left.
Rancid: Characteristic flavor common to autooxidized olive oil. This flavor is unpleasant and cannot be corrected.
Ripely fruity: Flavor of olive oil obtained from ripe fruit. It generally has a somewhat flat odor and a sweet taste.
Rough: Characteristic perception in certain oils. They produce a thick, pasty mouth sensation when they are tasted.
Soapy: Flavor that is reminiscent of that produced by green soap.
Sweet: Pleasant taste, not exactly sugary, found in olive oils in which the bitter, astringent and pungent attributes do not predominate.
Vegetablc water: Characteristic flavor acquired by the oil as a result of poor decantation and prolonged contact with vegetable water.
Winey-Vinegary: Characteristic flavor of certain oils reminiscent of wine or vinegar. It is mainly due to the formation of acetic acid, ethyl acetate and ethanol in large amounts.

Sensory evaluation of virgin olive oil has been proposed by IOOC (20). According to IOOC any person should be able to participate in the panel team. The tasters should be able to answer a special questionnaire.

TABLE 14.4
"Aroma and Flavor" Detected in Olive Oil.*

Aroma and flavor characteristics of olive oil	
Apple	Sour/winey/Vinegary/acid
Other ripe fruit	Rough
Green (leaves, grass)	Metallic
Bitter	Mustiness/humidity
Pungent	Muddy sediment
Sweet	Rancid

*IOOC (20)

The booths for tasters should be sufficiently large and comfortable (Fig. 14.1). The test room will be designed to provide to the panel team suitable environment, which facilitates work and helps to improve the repeatability and reproducibility of the results.

Table 14.4 gives the "aroma and flavor" detected in olive oil, while Table 14.5 shows a grading sheet for a sensory evaluation of olive oil.

For a "global" quality evaluation of virgin olive oil the IOOC proposed, in 1989, the following equation.

O.Q.I. = $2.55 + 0.91SE - 0.78AV - 7.35\ K_{270} - 0.066\ PV$
(Over-all, Quality
Index)

Where,
SE: Sensory evaluation (from 3.5 - 9.0)
AV: Acid value (from 0.1-3.3)
K 270: Absorbance at 270nm (from 0.08 - 0.25)
PV: Peroxide value (from 1.0 - 20.0)

TABLE 14.5

A Grading Sheet for a Sensory Evaluation of Olive Oil.*

Defects	Sensory Characteristics	"Overall" Evaluation (points)
None	Olive fruity Olive fruity and fruitiness of other fresh fruit	9 8 7
Slight and barely perceptible	Weak fruitiness of any type	6
Perceptible	Rather imperfect fruitiness, anomalous odors and tastes	5
Considerable, on the border of acceptability	Clearly imperfect, unpleasant odors and tastes	4
Great and/or serious, clearly perceptible	Totally inadmissible odors and tastes for consumption	3 2 1

*IOOC (20)

References

1. International Olive Oil Council (IOOC), International agreement on olive oil and table olives. United nations conference on trade and development. TD/Olive Oil. 8/7., Rome (1986).
2. International Olive Oil Council (IOOC), International trade standard applying to olive oils and oil-residue oils. COI/T.15/No 1., Madrid (1984).
3. International Olive Oil Council (IOOC), International trade standards applying to olive oils and olive residue oils. COI/T.15/NC No 1 Rev. 1., Madrid (1987).
4. Cucurachi, A., Final operation. In Olive Oil Technology. Moreno Martinez, J.M., Editor. FAO, Rome (1975).
5. FIAtron Process systems, On-line free fatty acid analysis. Bulletin, FIAche AOA, 1002. Oconomowoc, WI, U.S.A. (1989).
6. American Oil Chemist's Society (AOCS), Official and tentative methods. Third edition including additions and revision. Champaign, Illinois (1978).
7. Maurikos, P., D. Gegkou and G. Iliopoulos, Polarographic study of peroxides in olive oil. Prakt. Panellenion Chem. Syned. Athens, Greece. Chem Chron. 1:118 (1972).
8. Montefredine, A. and L. Luciano, Quality characteristic of virgin Italian olive oils. Proposal of a new classification of virgin olive oils. Boll. Lab. Chim. Prov. 19:784 (1968).
9. Bartolomeo, D. and R. Sergio, Physicochemical features and acidic composition of some meridional virgin olive oils. Riv. Ital. delle Sost. Grasse. 46:467 (1969).
10. Ninnis, L.N. and M.L. Ninni, Stabilite thermique de l' huile d' olive et sa prevision par son absorption dans l' ultra-violet A 232 et A 268nm. Rev. Fran. des Corps. Gras 15 (N.7):441 (1968).
11. Jimenez, O.J.M. and Gutierrez Conzalea-Quijiano, R., Packaging of olive oil in commercial type containers. III. Preservation in glass, tin, polyethylene, and PVC containers. Changes in absorptivity at 232 and 270nm. Grasas y Aceites 21:329 (1970).
12. Dugan, L.R., Lipids. In Principles of Food Science. Part I. Food Chemistry. Fennema, O., Editor. Marcel Dekker Inc., N.Y. (1976).
13. Gutierrez Gonzalea-Quijano, R. and A.V. Romero, Estudios sobre el enranciamiento del aceite de oliva. XI. Comparacion entre las diferences pruedas para la determinacion del grado de rancidez. Grasas y Aceites 11:67 (1960).
14. Casillo, R., Quality of virgin olive oils Thiobarbituric acid test. Riv. Ital. delle Sost. Grasse. 45:7535 (1968).
15. Kiritsakis, A., Quality studies on olive oil. Ph.D. Dissertation, Michigan State University. East Lansing Michigan, U.S.A. (1982).
16. Koutsaftakis, A., N. Psillakis, A. Kiritsakis, L. Mikros and D. Vamvoukas, Comparative trials of olive oils extraction systems. Report for the period

1978-1979. Publication of Olive and Subtrop. Plants Inst., Chanea, Greece (In Greek) (1979).
17. Carocci, C., Ancora Complesso Olearia "Sinolea" per l' estrazione dell'olici della olive. Instit. Sper. Per.l' Oliv. e l' Oleif. Imperia (1963).
18. Petruccioli, G., Prove Technologhe sul complesso "Sinolea-Pressa" e sulla centrifuga-Wesfalia. Ind. Agrarie 3:377 (1965).
19. Frezzotti, G. and M. Manni, Olive oil processing in rural mills. FAO, Rome. Agricultural development paper No. 58 (1956).
20. International Olive Oil Council (IOOC), Organoleptic assessment of virgin olive oil. COI/T.20/Doc.No. 3., Madrid (1987).

Chapter 15

Olive Oil Adulteration

Olive oil, which is a natural fruit product of fine aroma and pleasant taste, is also known for its nutritional and health value. Because of these qualities, olive oil commands a greater price in the international market than other vegetable oils. Consequently adulteration of olive oil with cheaper oils is a temptation. Some cases of adulteration are hazardous to the public's health.

Oils known to be widely used for adulteration include: olive pomace oil, corn oil, peanut oil, cottonseed oil, sunflower oil, soybean oil and poppy seed oil. In addition castor oil, pork fat (lard), as well as other animal fats have been occasionally used in small quantities. Adulteration with reesterified oils, and with industrial grade colza (rapeseed oil) has also been reported.

Firestone (1) reported that many brands of olive oil available in the USA market, contained undeclared esterified olive oils, pomace olive oils and seed oils as well. Based on these findings the Food and Drug Administration (FDA) started a program to control olive oil adulteration and mislabelling, which resulted in a decrease of undeclared esterified olive oils from 65% to 13% (2).

Methods and Techniques for Detecting Adulteration

Several physical and chemical tests have been proposed for detecting adulteration of olive oil (3-7). Table 15.1 shows the changes on certain constants (refractive index, specific weight, iodine value) of adulterated olive oil. Adulteration also changes the squalene content. Olive oil contains squalene which ranges from 136-708 mg/100g of oil (see Table 4.9). Any value outside these limits indicates adulteration of olive oil (8).

UV absorbance is often used for detecting olive oil adulteration. Olive oil absorbs 3-4 times less in the region 208-210nm than other vegetable oils. Ninnis and Ninni (5) proposed a quantitative method for determining olive oil adulteration based on the absorbance at two wave lengths (210, 268nm).

Gas chromatography has also been used for estimating the admixture of olive oil with vegetable oils (9,10). A combination of gas and liquid

TABLE 15.1

Changes of Certain Constants in Adulterated Olive Oil.

	Constants		
Oil	Refractive index	Specific weight	Iodine value
Cottonseed	+	+	+
Peanut	+	No change	+
Tea seed	No change	No change	No change
Soybean	+	+	+
Sunflower	+	+	+
Corn	+	+	+
Rape seed	+	Slight change	+

chromatography permits the detection of as little as 5% of a foreign oil (11).

Tsimidou *et al.* (12) used reversed phase HPLC to investigate the liquid fraction of oils obtained after low temperature crystallization and suggested that this procedure may be applied to assist the analysis for detecting olive oil adulteration. Casadei (13) used also HPLC to detect adulterated and misbranded olive oil products with hazel-nut oil. According to him, HPLC is appropriate to detect olive oil adulterated with esterified oil.

One of the criteria proposed by the IOOC (4) for olive oil purity, refers to saturated fatty acid content at position-2 of the triglyceride. Reesterified oils differ in their fatty acids composition at position-2. Hydrolysis by means of pancreatic lipase, which specifically attacks position-2, allows the differentiation between virgin and esterified olive oils.

The saturated fatty acid level (sum of palmitic and stearic acids) at position-2 should not exceed 1.5% in virgin olive oil, 1.8% in refined olive oil or in pure olive oil, and 2.2% in refined olive-pomace oil (4). Paganuzzi (14,15) reported that quantitation of erythrodiol and uvaol provides a good indication for differentiation between pressed and solvent extracted olive oils.

Because soybean oil is much richer in γ-and δ-tocopherols and cottonseed oil in γ-tocopherol than olive oil (Table 4.13), quantitative determination of these tocopherols has been suggested as a method for determining the adulteration of olive oil with the less expensive soybean or cottonseed oils (16,17).

At times, reesterified olive pomace oils are used to adulterate. The detection of elaidic acid, by IR spectroscopy, in such oils reveals the adulteration (18). Analysis of the entire unsaponifiable fraction (eg sterols, tocopherols, triterpenic alcohols) will probably provide additional tests for sharpening the differentiation among vegetable oils (6,8,18).

The determination of β-sitosterol content in olive oil as well as the value of the ratio:

$$\frac{\beta\text{-sitosterol}}{\text{campesterol} + \text{stigmasterol}}$$

is used for detecting olive oil adulteration with seed oils (18). Under standard conditions, virgin olive oil has a β-sitosterol content of 94% or higher, this substance being associated with stigmasterol and campesterol.

According to Gutfinger and Letan (17), large quantities of stigmasterol indicate the presence of soybean oil in olive oil. Sterols are converted into their trimethylsilylether (TMS) derivatives and are analyzed with GLC in SE-30 or JXP (3%) (6).

Solvent extracted oils may be distinguished from oils obtained by pressure process, applying simple procedures such as: Bellier, Carocci-Buzi and Vizern test, or determining the erythrodiol content by GLC (6). Chapter 16 provides more information on detecting olive oil adulteration.

References

1. Firestone, D., Control of olive oil adulteration and misbranding in the United States. Riv. Ital. delle Sost. Grasse. 64:293 (1987).
2. Firestone, D., K.L. Carson and R.J. Reina, Update on control of olive oil adulteration and misbranding in the United States. J. Am. Oil Chem. Soc. 65:788 (1988).
3. International Olive Oil Council (IOOC). Trade standards for virgin olive oils. COI No 2/66/15-II (1966).
4. International Olive Oil Council (IOOC), International trade standards applying to olive oils and olive-residue oils. COI/T.15/No 1 (1984).
5. Ninnis, L.N. and M.L. Ninni, L'importance pour l'analyse de l'huile d'olive de l'absoption dans l'ultra-violet (190 a 220nm). Rev. Franc. des Corps Gras. 13 Anne 1 (1966).
6. Fedeli, E., Lipids of olives. Prog. Chem. Fats and other Lipids. 15:57 (1977).
7. Codex Alimentarius Commission WHO/FAO, Recommended International standard for olive oil, virgin and refined and for refined olive residue oil. CAC/RS-33 (1970).

8. Cuisa, W. and A. Morgante, Polycyclic aromatic hydrocarbons in olives. Quad. Merced. 13:31 (1974).
9. Colakoglu, M., Donnes analytiques nourelles sur les huiles d'olive. Rev. Franc. des Coprs Gras. 13 Anne 4:261 (1966).
10. Gegiou, D. and M. Georgouli, Detection of reesterified oils. Determination of fatty acids at position-2 in glycerides of oils. J. Am. Oil Chem. Soc. 57:313 (1980).
11. Kapoulas, V.M. and S. Passaloglou-Emmanoulidou, Detection of adulteration of olive oil with seed oils by a combination of column and gas liquid chromatography. J. Am. Oil Chem. Soc. 58:694 (1981).
12. Tsimidou, M., R. Macrae and I. Wilson, Authentication of virgin olive oils using principal components analysis of triglyceride and fatty acid profiles. Food Chemistry. 25:227 (1987).
13. Casadei, E., First results on detection of adulterated olive oil products with hazel-nut and/or esterified oils by HPLC of triglycerides. Riv. Ital. delle Sost. Grasse. 64:373 (1987).
14. Paganuzzi, V., Distribution of alcoholic components in the unsaponifiable fraction of olive drupes. IV. 4-methyl sterols. Riv. Ital. delle Sost. Grasse. 57:291 (1980).
15. Paganuzzi, V., Influence of origin and conservation state on the triterpene dialcohols relative content of untreated olive oils. III. Pressed oils from Greece. Fette Seifen Anstrich. 84:115 (1981).
16. Tiscornia, E. and G. Bertini, Recent analytical data in chemical composition and structure of olive oil. Riv. Ital. delle Sost. Grasse. 49:3 (1972).
17. Gutfinger, J. and A. Letan, Studies of unsaponifiables in several vegetable oils. Lipids. 9:658 (1974).
18. Pallotta, U., Analytic problems in the ascertanment of olive oil genuinesses. Lipids Vol. 2 Technology. Paoletti, R., G. Jacini and R. Porsellati, Editors. Raven Press, New York (1976).

Chapter 16

Determinations on Olive Oil-Adulteration Tests

Methods used for determining acidity and oxidation of olive oil and evaluating its sensory characteristics has been described in Chapter 14. Some methods used for the determination of certain constituents and impurities on olive oil and other detecting its adulteration are presented in this chapter.

Chlorophyll Determination

The chlorophyll content of olive oil (responsible for its unique green color) can be determined by the AOCS method (1), measuring the absorbance at 630, 670 and 710nm, by a spectrophotometer. The value is calculated using a formula.

Rahmani and Csallany (2) proposed a rapid and precise HPLC method for determining the chlorophylls a and b and pheophytins a and b in olive oil, using a column of μ-porosil. Determinations are made at 430nm for chlorophyll a, 452nm for chlorophyll b, 409nm for pheophytin a, and 432nm for pheophytin b. The method is precise at 0.01 μg of chlorophyll a or b and at 0.02 μg of pheophytin a or b.

Determination of Phenols

Phenols of olive oil may be determined applying the method based on color formation using Folin Ciocalteau as a reagent (3,4). The absorbance of the blue color formed is measured at 725nm. For the quantitative analysis caffeic acid can serve as a standard.

Moisture Determination

The moisture content of olive oil can be determined by drying it in an oven, by the distillation procedure (1) or by infrared (IR) radiation. Moisture determination, by IR, is carried out on a special balance furnished with a lamp. The weight loss, during evaporation, is calculated by a balance system and it is expressed in a percentage of the sample weight.

Determination of Soap Content (CAC method, recommended by IOOC)

This method is based on volumetric titration with hydrochloric acid, using bromophenol blue as indicator (5). The dissolved soap is expressed as sodium oleate (%).

Cold Test (AOCS method)

The cold test measures the resistance of olive oil and other oils to crystallization and is commonly used as an index of evaluating the winterizing or similar stearin removing processes (see Chapter 9).

Bottles containing the samples are immersed in an ice and water bath so that the entire bottle is submerged. The temperature should remain at 0°C. To pass the test the sample must be clear, limpid, and brilliant at the end of 5½ hours (1).

Adulteration and Genuineness Tests of Olive Oil

Tests Based on Color Formation

Nitric Acid Test (Synodinou-Konsta method). One of the first tests for detecting adulteration of olive oil, is that of Hauchecorne based on color formation using nitric acid. The initial Hauchecorne method was later modified by the Greeks Synodinou and Konsta (6-10). Any color formed, other than hey to hey-yellow indicates the presence of seed oil or refined olive oil or olive pomace oil in the olive oil. Substances responsible for the color development may be formed and as a result of autoxidation. Therefore this method should not be used in oxidized olive oils (8).

Halphen Test. The Halphen test is used to detect the adulteration of olive oil with cottonseed oil (7-9). It is characteristic of the cyclopropenoic acids, malvic and stercoulic, found in the cottonseed oil. A Halphen reagent containing equal volumes of amyl alcohol and sulfur solution, 1% in CS_2, is used. The formation of red color indicates the presence of cottonseed oil. The depth of color is to a certain extent proportional to the amount of cottonseed oil in the olive oil.

Sesame Oil Tests. There are different tests (Baudouin, Villavecchia, Fabris), commonly used for the detection of sesame oil in olive oil (5,7). They are based on the color formation due to the presence of the phenolic substances, sesamol, sesamolin (Fig. 16.1) and sesamin on the sesame oil (11). The formation of crimson color (Baudouin) yellow green (Fabris) and any color (Villavecchia) in the lower layer of the oil sample

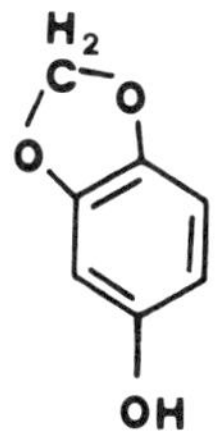

Sesamol

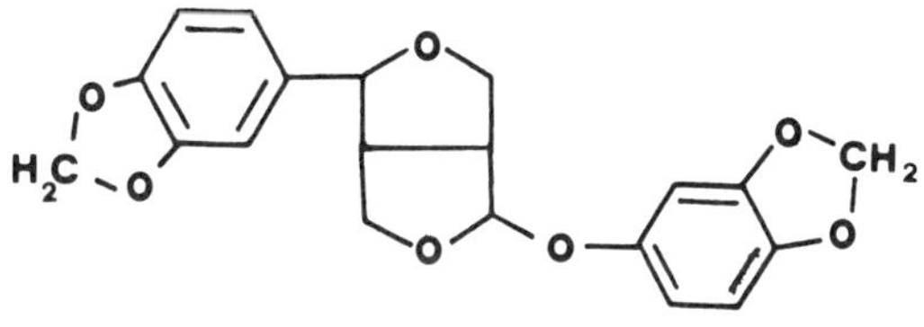

Sesamolin

Fig. 16.1. Structures of sesamol and sesamolin.

indicates the presence of sesame oil (11). Refined sesame oil does not always give positive reaction, by applying the Villavecchia test.

Tests Based on Residue Formation

Carocci-Buzzi test. The Carocci-Buzzi test is detecting adulteration of olive oil with olive pomace oil (5,7). Flake formation is an indication of adulteration. Clouding with no flake formation does not necessarily prove absence of oil pomace oil from the olive oil.

Vizern-Guillot test (CAC method, recommended by IOOC). The Vizern-Guillot test is detecting the adulteration of virgin and refined olive oil with semi-drying oils (Iodine Value 100-150) (5,12-15). All seed oils fall into this category. The method is based on the reaction between the semi-drying oils and the reagent bromine. If semi-drying oil is present a flocculent precipitate will form varying in quantity according to the percentage of adulteration and the nature of the adulterating oil. The Vizern-Guillot test is not considered reliable for olive oils with high iodine value (13).

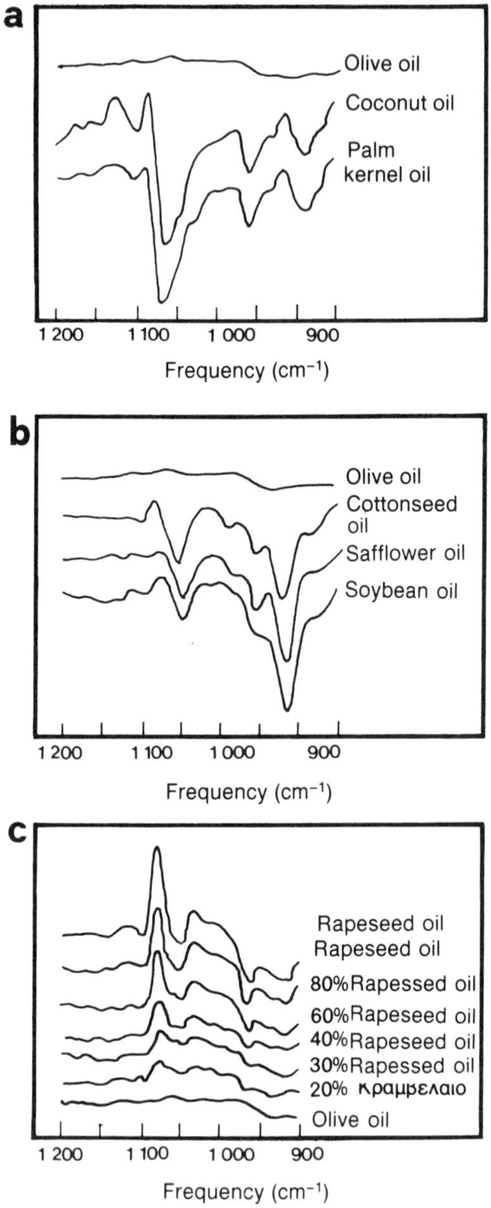

Fig. 16.2. Infra-red (IR) spectra of virgin and adulterated olive oil (11).

Other Tests For Determining The Genuineness of Olive Oil

(CAC method, recommended by IOOC)

Bellier-Marcille Index. The Bellier-Marcille index is used for detecting adulteration of olive oil, mainly with peanut oil (5,13). It is the temperature (°C) at which precipitation (appearance of cloudiness) of the salts of the oil fatty acids commences. The Bellier-Marcille index for virgin and refined olive oil should not be higher than 17 (see Chapter 14).

Detection of Olive Oil Adulteration by UV Lamp

An ultraviolet (UV) lamp, known as a wood lamp, is used for detecting adulteration of olive oil with refined olive oil, or refined olive pomace oil (7). Another UV lamp could also be used.

Fluorescent light, yellow to orange-yellow color, is characteristic for virgin olive oils. Appearance of blue fluorescent color indicates the presence of refined olive oil or olive pomace oil in the tested oil.

Detection of Olive Oil Purity by IR Spectrophotometry

IR spectrophotometry can be used for detecting the purity of olive oil (11,16). The different oils show a characteristic absorbance in the IR region. Figure 16.2 presents some IR spectra of olive oil to which certain seed oils were added at different quantities.

References

1. American Oil Chemist's Society (AOCS), Official and Tentative Methods. Third edition including additions and revision. Champaign, Illinois. (1978).
2. Rahmani, M. and A.S. Csallany, Mise au point d' une methode de chromatographie liquid a'haute performance (CLHP) pour la determination des pigments chlorophylliens dans les huiles vegetales. Revue Francaise des Coprs Gras. 32 annee. n 6-7:1 (1985).
3. Gutfinger, J., Polyphenols in olive oils. J. Am. Oil Chem. Soc. 58:966 (1981).
4. Kiritsakis, A., Quality studies on olive oil. Ph.D. Dissertation, Michigan State University. East Lansing, Michigan, U.S.A. (1982).
5. Codex Alimentarius Commission (CAC), FAO/WHO, Methods of analysis for edible fats and oils. CAC/RS 33-1970, Rome (1969).
6. Balatsouras, G., Olive Oil, seed oils, lipids. Karamperopoulos, Editor. Athens, Greece (In Greek) (1986).

7. Chemical Greek Laboratory, Code of foods, drinks and other products. Part B. Formal methods of testing foods and drinks. National Publishing Office. Athens, Greece (In Greek) (1976).
8. Demopoulos, I., Lab Notes for Food Chemistry. Ministr. of Education. Athens, Greece (in Greek) (1982).
9. Demoulas, K., Quality Control of Vegetable Foods. Ministry of Education. Athens, Greec (In Greek) (1981).
10. Galanos, S., Chemistry of foods and drinks. General methods of analysis. Papazisis, Editor. Athens, Greece (In Greek) (1947).
11. Mehlenbacher, V.C., The analysis of fats and oils. The Garrard Press Publishers. Champaign, Illinois (1960).
12. European Economic Community, Analysis of olive oil. Bollettino dei Chimici dei Laboratori Provinciali.3:246 (1977).
13. International Olive Oil Council (IOOC), Common methods for analysis of olive oils. T.14/DOC. No4/Corr.1 Annex III, Madrid, Spain (1966).
14. IUPAC, Standard Methods of the Oils and Fats Section, 5th edition. Butterworths, London, England (1966).
15. Martel, J., Fractional crystallization of the triglycerides of virgin olive oil. Presence of semi-drying oils. Grasas y Aceites. 28:189 (1977).
16. Kiritsakis, A., Lectures for Fats and Oils Analysis. Technological Educational Institute (TEI). Thessoloniki, Greece (In Greek) (1984).

Chapter 17

Biological Effects of Olive Oil on Human Health

Anthony Kafatos and George Comas*

*University of Crete School of Medicine
Department of Social Medicine
Supported by a grant from the European Economic Community through the Mediterranean Integrated Programs and the Greek Ministry of Health

Olive oil, a frequently used mythological symbol of peace, strength, and virility has always been acknowledged as having a sacred nature. The ancients respected its medical and cosmetic value and awarded it religious status. Many ancient physicians such as Hippocrates and Galen recognized its therapeutic benefit and extolled it as a combined food and medicine. In recent years as nutritional data has accumulated, medical science is verifying what the doctors of antiquity already knew: that olive oil is a source of significant nutritional value in human health and a preventive factor in many diet-related illnesses.

Olive Oil as a Source of Dietary Lipids

Lipids are important dietary macro nutrients that supply cells with the chemical energy stored in their abundant C-H bonds. In addition, lipids are important in the functioning of many cell organelles; a deficiency of linoleic acid, a dietary fatty acid, can induce a skin disorder and growth disturbances. Lipids are also a source of insoluble vitamins; they play a role in the growth and development of the reproductive and nervous system, as well as the skin and the liver (1). Among the essential fatty acids that the body cannot synthesize are linoleic acid (18:2,ω-6) and α-linolenic acid (18:3,ω-3). Their functions are to regulate fluidity and permeability in the cell membrane structure and to serve as precursors in prostaglandin synthesis. Besides a deficiency of these essential fatty acids having serious clinical implications, large quantities are not necessary and may be harmful.

It is recommended that 1% of the total energy intake must be provided from essential fatty acids to meet physiological needs. Because polyunsaturated fatty acids (PUFAs) have also been shown to reduce the level of plasma LDL cholesterol and consequently to lower risk of atherosclerosis many nutritionists have advocated a greater intake of PUFAs as a protective factor against lipid metabolism disorders such as hypercholesterolemia. The negative consequence of this dietary approach is the risk of lipid peroxidation. Peroxidation less common in saturated and monounsaturated fatty acids is very rapid with polyunsaturates because vulnerability to peroxidation increases with the number of double bonds. The dilemma of polyunsaturated fatty acid composition is that excessive intake of these unstable substances can be almost as harmful as a deficiency. Lipid peroxidation results in a chain reaction that produces free radicals which can have toxic intracellular effects (1).

Free radicals are biologically unstable and extremely reactive because they contain an unpaired electron in their electrochemical configuration. An accumulation of free radicals leads to the formation of toxic hydroperoxides, which also can initiate more free radicals forming a chain reaction. Disorders related to free radical formation include liver diseases (2) inflammation (3) malignancies such as large bowel cancer (4-6), and accelerated aging (7-9). In addition, PUFA peroxidation can contribute to atherosclerosis by inducing platelet hyperaggregation and aggravating endothelial lesions (1). Moreover the fatty acids released from atherosclerotic lesions are peroxidized, a process leading to localized decrease in prostacyclin. The reduced prostacyclin in the fibrous plaques and in the lipid striae accelerates the formation of atheromatous plaque (1).

The body is protected from the toxic effects of these reactive compounds by certain antioxidant defense agents such as the tocopherols. Enclosed within the phospholipid bilayer of the cell membrane, the tocopherols interact with the PUFAs, mitigating the peroxidative processes and thus reducing free radical formation. An important facet of this problem is that normal dietary consumption of tocopherols is sufficient to defend the body from an excessive intake of polyunsaturates. Tocopherols occur naturally in the α, β, γ and δ forms. The α-tocopherol is the most biologically active, while the β and γ forms have limited activity and the δ-tocopherols are inactive. Olive oil contains α-tocopherol. Wheat germ oil, containing only γ-tocopherols and δ-tocopherols has little or no protective property. Futhermore, requisite vitamin E intake is increased 200 times the RDA level of 10 mg when PUFA consumption rises (9). In addition when vitamin E is consumed its absorption is reduced by one third in the presence of polyunsaturated linoleic acid (10). Thus overconsumption of polyunsaturated fatty acids carries the risk of vitamin E deficiency and the accompanying effects.

In this regard olive oil's chemical composition makes it ideally suited as a nutritional defense against lipid peroxidation. As noted above olive oil is high in monounsaturated oleic acid (18:1ω-9) and carries a lower risk of peroxidation than polyunsaturated oils, while having a high content of biologically active α-tocopherol hence its enhanced antioxidative stability (1). Olive oil's optimal tocopherol/polyunsaturated ratio of more than 1:8 (mg of vit E per g of linoleic acid) allows it to supplement the required daily essential fatty acids while protecting the body from vitamin E deficiency (10). Thus the ideal dietary approach would be to replace saturated animal fat consumption with a higher intake of monounsaturated rich olive oil. Indeed, Keys (11) has suggested that the monounsaturates to saturates ratio as opposed to the polyunsaturates to saturates ratio also be considered as a nutritional parameter in the epidemiological study of coronary heart disease incidence. Consequently when all factors are considered olive oil has an optimal proportion of essential fatty acids.

Olive Oil and the Stomach

Nearly a century ago, Ewald and Boas (12) observed that supplementing gruel meal with olive oil reduces gastric fluid concentration. In 1926 Farrel and Ivy (13) conducted a series of studies confirming Ewald and Boas data. They speculated although with little hard data, that this response was due to the stimulation of a gastrointestinal hormone enterogastrone. Castelli et al. also reported that consuming olive oil relieved abdominal pain and lessened indigestion after a meal. In an experiment with rats, Benvestito *et al.* (14) noted that because it enhanced alkaline secretion by the mucosal cells lining the stomach, olive oil added to meal mitigated the development of peptic ulcers. In addition, Taits (15) verified olive oil's therapeutic effects on peptic ulcers when he observed a decline in stomach ulcers when olive oil was substituted for saturated animal fats in the diet.

Recently, however, Charbonnier (16) published a study on gastric lipid tolerance. Olive oil was found to have the mildest effect on the lower esophageal sphincter (LOS) in comparison to polyunsaturated and saturated fatty acids. On the other hand butter induced a significant protracted drop in LOS tonus and a pronounced lengthening of gastric semi-emptying time and caused the most gastric irritation. Polyunsaturated sunflower oil yielded moderate results less than those of olive oil. Although olive oil cannot replace clinical medication in the treatment of gastroduodenal ulcers it clearly has a protective therapeutic effect on gastric lesions.

Olive Oil and the Intestine

Olive oil also helps to alleviate atonic constipation. By inducing the secretion of cholecystokinin which activates gall bladder contractions and stimulates peristalsis of the small intestine, olive oil has a gentle laxative effect which aids in relieving mild chronic constipation (17).

Olive Oil and the Bile Ducts

Olive oil has a relaxing positive effect on gall bladder contractions because it activates the hormone cholecystokinin. It enlarges the sphincter of Oddi and keeps it open for a protracted time while it stimulates evacuation of bile (1). Olive oil's beneficial cholagogic effect has been the subject of much experimental research. In a study on the dietary treatment of cholecystopathies, Demole (18) reported that among lipids, the best cholecystokinetic results were obtained with olive oil. It was concluded that olive oil was the most effective cholagogue in the treatment of biliary dyskinesia because it gently stimulated steady bile drainage and relaxed the sphincter of Oddi. Recently Charbonnier (16) reproduced Demole's results using duodenal intubation. He recorded that because of olive oil's apparent mild, yet vigorous cholecystokinetic effect, as well as its suppression of hepatic bile secretion during gall bladder drainage (found by measuring bromosulphophthalein in the bile) it was a unique and complete cholagogue. In addition olive oil enhances the detoxicative properties of the liver by clearing bromosulphophthalein from the plasma.

Cholesterol gallstones are a major disease associated with dietary lipid intake. Like atherosclerosis, the development of cholesterol gallstones is most common in affluent countries with high living standards (19,20). Diets rich in saturated fats (21,22) cause a rise in biliary excretion of cholesterol and a decrease in lecithin and bile acids, which may be related to the pathogenesis of cholelithiasis. In addition to high cholesterol and high saturated fatty acid intake, it has been reported that a deficiency of adequate caloric intake and dietary lipids (23,24) as well as high PUFA consumption (25-27) are lithogenic. Indeed studies by Bucko and Ginter (28) but not by Borgman and Haseldem (29) or Schuller *et al.* (30) have verified that excessive PUFA in the diet may contribute to high bile lithogenicity. In contrast, Messini and Cairella (31) observed that olive oil consumption and the development of cholesterol gallstones were related inversely in many areas in Italy.

In summary, it can be concluded that diet is a key factor in the etiology and treatment of cholelithiasis. An ideal diet should include a balanced caloric intake of lipids and carbohydrates and should be low in dietary cholesterol and fatty acids, either saturated or polyunsaturated. Thus because of its cholesterol lowering effect, olive oil may contain the ideal proportions of fatty acids and cholagogic properties to make it well suited for the dietary therapy of cholelithiasis.

Olive Oil and Growth

Dietary lipids constitute 50% of the calories in the mother's milk and are an important constituent of infant nutrition. PUFA's such as linoleic and linolenic acid are metabolically essential for infants and continue to be necessary for the later growth and development of children. Indeed deficiency of essential fatty acids in infants has been linked to growth retardation and to liver skin and metabolic dysfunctions. The linoleic (18:2,ω-6) / to a-linolenic (18:3,ω-3) ratio however is just as important as the quantity. Deficiency of these important fatty acids may damage the development of the nervous system by inhibiting the production of such compounds as eicosapentaenoic (20:5,ω-3) and docosahexaenoic (22:6,ω-3) acids, which are essential structural components of brain tissue (32). In a study conducted on growing rats fed diets of olive oil, sunflower oil, and animal fat, Galli (33) observed inferior development of cerebral and hepatic tissue in the rats fed with the saturated or polyunsaturated lipid diets as compared to the rats that consumed olive oil. Thus it was demonstrated that olive oil optimal linoleic/2-linolenic ratio matches mother's milk most closely and is superior lipid source for growing children.

Another important consideration is that infants require PUFAs but a high intake of these fats may lead to serious vitamin E deficiency. Developing infants also are especially susceptible to PUFA peroxidation, which is provoked by vitamin E deficiency. The effect of a diet excessive in PUFA and deficient in vitamin E in infants has been shown to result in hemolytic anemia.

Laval-Jeantent *et al.* (34) have published a study conducted on rats on the effects of various dietary lipids on growth and development. They concluded that with regard to bone mineralization, olive oil produced better results than pure triolein. This result is probably due to the essential fatty acid content of the olive oil; the high level of antioxidants protected the bone tissue from peroxidation and supported growth and development in the children.

Olive Oil and Aging

One of the most important factors in the slowing of the biological clock is diet. Just as a nutritionally balanced diet is essential in the proper growth and development of young children, it is equally important in maintaining the health and well being of elderly people. One of the probable causes of the degeneration of the biological activity of cells is the accumulation of free radicals. As mentioned previously, free radicals are the products of chain reactions initiated by the peroxidation of PUFAs. Peroxidized free radicals accumulate within the cells, damaging the cells ability to perform many functions (35). They may fragment and deform nucleotide chains thus causing mutations distorting the membrane phospholipids and damaging the cell membrane. The permeability of brain cell membranes to water may be altered resulting in distention of the neurons and causing or aggravating signs of senile cerebral psychoorganic syndrome and/or dementia (36). Experiments conducted on a number of animals have confirmed that diets deficient in antioxidants (vitamin E), increase peroxidation and free radical formation while diets high in vitamin E inhibit peroxidation and extend life expectancy (7,37,38).

Harman (39) has demonstrated experimentally the deleterious aging effects of excessive PUFA intake. Animals consuming high amounts of PUFAs (over 2% of total calories from linoleic acid and over 0.5% from a-linoleic acid) developed premature senile dementia and deteriorated mental function. Moreover, mice administered polyunsaturated corn oil had a shorter life expectancy than mice consuming an olive oil diet high in vitamin E (40).

Laval-Jeantent *et al.* (34) demonstrated olive oil's favorable effect to bone calcification and hypothesized that this effect is due to the importance of oleate as a structural lipid in bone formation. They concluded that olive oil's moderate but balanced content of PUFAs and high amounts of vitamin E make it an ideal source of dietary lipid in old age as well as during childhood development.

Olive Oil and Diabetes

Because of complications in the metabolism of lipids and glucose, diet has played an important role in the treatment of noninsulin-dependent diabetes mellitus (NIDDM). Dietary approaches in recent years, however have encountered much controversy, they have shifted between support for a low-fat, high carbohydrate diet and advocacy of high unsaturated fat intake. Since diabetics are vulnerable to hyperglycemia and coronary heart disease, the correct dietary approach must moderate plasma glucose, insulin, and plasma lipids levels. Parameters that must be considered in evaluating dietary treatment include glycemic levels, plasma triglycerides and individual lipoprotein fractions (very low density lipoproteins, VLDL, low density lipoproteins, LDL; high-density lipoproteins HDL).

Garg *et al.* (41) recently published a study comparing the effects of a diet high in complex carbohydrates (25% fat, 60% carbohydrate) with a diet high in monounsaturated fatty acids (MUFA) (50% fat, 35% carbohydrate) on glycemic control and levels of plasma lipids and lipoproteins in patients suffering from NIDDM. The high MUFA diet reduced mean plasma glucose levels (and thus insulin requirements), decreased VLDL cholesterol by 35% and plasma triglycerides by 25%, and raised HDL cholesterol by 13%, in comparison to the high-carbohydrate diet.

Low-fat, high carbohydrate diets are recommended for diabetics in order to protect them from coronary heart disease by reducing the atherogenous LDL cholesterol. Clinical studies (42,43) however have revealed that MUFA diets produce equal reductions in LDL cholesterol. In addition excessive carbohydrate intake decreases glycemic control (44,45) and raises triglyceride levels (46,47) a particularly harmful combination for hyperglycemic patients with NIDDM.

These studies demonstrated that the monounsaturated diet improved glucose tolerance and had a positive effect on HDL, LDL, and triglyceride levels (48). The authors concluded that because of beneficial effects of the monounsaturated diet on HDL, LDL, triglyceride levels and glucose tolerance, monounsaturated fatty acids (olive oil) should play a role in the dietary treatment of NIDDM.

Olive Oil and Atherosclerosis

Atherosclerosis, the hardening of the arteries, is the most prevalent cause of death in industrialized countries. Its numerous related risk factors, although well studied, are ambiguous and complicated. Many patients have a genetic tendency toward this coronary disease but a number of risk factors related to behavior such as smoking, lack of physical activity, hypercholesterolemia, and obesity as well as arterial hypertension have been postulated to promote athero-sclerotic plaque formation. Additional factors include sex (male), age (44-54), diabetes, gout, hypertriglyceridemia, and type A behavior.

In the last 30 years many researchers have undertaken numerous studies, both clinical and epidemiological, to investigate the relationships between diet and the occurrence of atherosclerosis. It has been observed that the incidence of this disease is higher in affluent than in developing societies, where the standard of living is lower. Population studies from different geographic regions, where varying eating habits prevail have reported pronounced differences in incidence rates of morbidity and mortality due to coronary disease.

One of the most important of these populations studies is the seven country study (49). This study is a landmark documentation of the relationship between extensive consumption of animal fat and the onset of cardiovascular disease. By comparing diets of regional populations with the incidence of coronary heart disease (CHD), the study has revealed an etiopathogenic correlation between diet type and the incidence of atherosclerosis. It has been demonstrated that animal fats rich in saturated fatty acids, increase plasma cholesterol which precipitates atherosclerotic lesions (50,51). Saturated fatty acids prevalent in Western diets are mostly responsible for the cholesterol-raising effect of saturated animal fats (52). In contrast, polyunsaturated fatty acids such as linoleic acid found in vegetable oils have a cholesterol-reducing effect and are regarded as beneficial, anti-atherogenous dietary lipids. Thus many nutritionists have recommended diets high in polyunsaturated fatty acids in the treatment of hyperlipidemia. Olive oil, rich in monounsaturated oleic acid (18:1) has not been reported either to increase or to decrease serum cholesterol levels.

Cholesterol is the most common sterol found in man. It is not only the precursor of steroid hormones and bile acids but also an important component of the cell membrane. Cholesterol regulates the fluidity of the cell membrane lowering it in the presence of unsaturated fatty acids and raising it where saturated fatty acids are abundant (1). Lipids, however, are insoluble in water; thus when fats in the form of triglycerides and cholesterol are absorbed by the intestine, they must be carried in the bloodstream bound to large lipoproteinic compounds called chylomicrons. Triglycerides are transported to the adipose tissue, while cholesterol is deposited in the liver.

The liver plays a key role in the schema of plasma lipid transport by regulating the exogenous and the endogenous plasma cholesterol levels. Triglycerides and cholesterol stored in the liver as well as originally synthesized lipids, are dispensed into the bloodstream attached to very-low density lipoproteins. Like the chylomicrons, the VLDLs transport the triglycerides to the adipose tissue and then turn into low-density lipoproteins (LDL) which distribute cholesterol to the surrounding cell tissue. Specialized membrane receptors recognize the apoprotein B, a protein component of LDL, and the cells take the cholesterol for their metabolic requirements. Thus if there is an imbalance in cholesterol inhibition, caused by insufficient membrane receptors or by excessive LDL-C intake, there is a subsequent increase in plasma LDL an atherogenous risk factor.

On the other hand, high-density lipoproteins remove cholesterol from the cells after it has completed its biological function. The cholesterol is esterified and incorporated into the HDL with the assistance of the enzyme L-CAT, which is activated by apoprotein A, a component of HDL. HDLs transfer the cholesterol to the liver, where it is excreted via the bile ducts. Thus an increase of apoprotein A in plasma HDL is regarded as beneficial, protecting against atherosclerosis. Harvengt (53) and Castelli *et al.* (54) have verified that high HDL levels are linked with lower atherosclerosis risk and longer life expectancy.

In this context it is incorrect to assert that monounsaturated fatty acids have an "indifferent" effect on plasma cholesterol. Numerous studies have confirmed that MUFAS increase HDL-C, while PUFAS appear to lower the level of this protective antiatherogenous compound (55,56). In evaluating clinical treatment for hyperlipidemia it is important to note that the cholesterol suppressing property of PUFAS is much weaker than the cholesterol-raising effect of SFAs. Roughly 2gm of PUFAS are necessary to counterbalance 1 gm of saturated fat intake (52).

Thus in the dietary management of lipid metabolism disorders such as hypercholesterolemia it is fundamental to begin by curtailing or eliminating saturated fat intake before increasing unsaturated fat consumption. In fact, the improvement in plasma tool cholesterol when saturated fatty acids were cut from the diet was twice as great as the improvement obtained when an equal quantity of dietary polyunsaturates were added. When saturated fats are replaced by a monounsaturated fat such as olive oil, the effect of plasma total cholesterol is equal to results obtained when saturated fats are reduced. Ferro-Luzzi *et al.* (57) confirmed the positive results of substituting olive oil for saturated animal fats in a study conducted in rural southern Italy. The subjects examined were from the industrial regions of Italy and still consumed the traditional Mediterranean diet, high in olive oil. The researchers, in an attempt to stimulate the Western diet, replaced olive oil with saturated animal fats in the diets of 25 men and 23 women. This diet resulted in a pronounced rise in total cholesterol and LDL cholesterol in subjects and in an additional significant increase in HDL cholesterol in the women. After 6 weeks, the participants reverted to their initial values. This study corroborated the favorable results that can be achieved by replacing saturated fats in the diet with monounsaturate rich oils such as olive oil, without adjusting polyunsaturate consumption. Clinical and epidemiological studies have demonstrated that MUFAs are as effective as PUFAs in reducing total cholesterol; they also have the unique property of increasing HDL levels. Moreover, monounsaturate-enriched diets seem to be correlated with low incidence of coronary heart disease. The results of Keys' seven-country study (58) agree with this assessment. A prospective study screening 10,000 men from the United States, Finland, Italy, Greece, Yugoslavia, Holland and Japan has contributed a wealth of material comparing established dietary habits with coronary mortality. The conclusions of this study, based on 10 year age standardized death rates per 10,000 men, disclosed greater coronary heart disease in Italy (111.3 per 1,000 men) than in Greece (72.5 per 1,000 men in Corfu and 6.3 in Crete) and a substantial difference between non-Mediterranean and Mediterranean countries in mortality from all causes and in CHD rates. A notable difference has been found between US railroad workers and Greeks in death rates (52).

Another major finding was that in the presence of equivalent plasma cholesterol levels, the risk of incurring coronary disease is approximately the same for Americans and for Finns, whereas it is significantly less for persons living in the Mediterranean area, where olive oil constitutes the main fatty substance in the diet (59).

Regarding defined eating habits and serum cholesterol levels Keys notes correlation between serum cholesterol and dietary intake of saturated fats. He summarizes the concerns of his epidemiological study as follows: Major items in the 15 year experience include demonstration that the risk of coronary death of middle-aged men is strongly related to the blood pressure, to the level of cholesterol in the blood serum and to smoking cigarettes. In regard to the diet, the average cholesterol in the blood serum was closely related to the average percentage of calories from saturated fatty acids in the diets of the populations. But the major ills of mankind seldom have only a single cause. Coronary heart disease the leading cause of death in many of the more prosperous populations is agreed to be multi-factorial in origin, meaning that a number of factors influence the development of the disease. Extreme caution is needed in attempting to make detailed interpretations of the findings in the seven-country study and in other epidemiological investigations. Why are the men in Greece and Italy less plagued by coronary heart disease than the men of Northern Europe and the US? It seems clear that the difference in the modes of life in Italy and Greece as compared with Northern Europe and the US makes a great difference in the susceptibility to CHD. The use of olive oil as a major part of the diet and as the predominant fat consumed in a high-fat diet is certainly compatible with an excellent record in the incidence of coronary heart disease and longevity.

In the Mid-1970s the risk factors for myocardial infarction had risen rapidly (60) in the rural and urban populations of Crete. The total serum cholesterol levels of middle-aged adult Cretans increased 25% over to cholesterol levels in the early 1960s. This change can be attributed to the progressive shift in the traditional diet, a diet which had remained uniform for nearly 4000 years (52). Indeed, olive oil intake accounted for 32.6% of total daily energy intake in 1960 but for only 21.7% in 1979 (59). Recent data confirm that olive oil constitutes 25% of the total energy intake in the average Cretan diet (49,60,61). This level of consumption is reflected in the adipose fatty acid composition. The analysis of subcutaneous fat has been used as a validation procedure for the long-term dietary fat intake. Christakis *et al.* (61) in 1962 and Kafatos *et al.* (60) in 1988 studied dietary intake and fatty acid composition in samples of Cretan population. The method of obtaining and analyzing the subcutaneous fat was the same (60,61).

TABLE 17.1

Adipose Tissue Fatty Acid Composition of Cretan Adult Males in 1962 and 1988

Fatty acids	Year 1962 (200 men) %	Year 1988 (250 men) %	Change %
Saturated F.A.	18.8	23.4	+24.5
Monounsaturated F.A.	68.7	66.1	- 4
Polyunsaturated F.A.	12.3	8.6	-30
P:S Ratio	0.65	0.36	-44.6
M:S Ratio	3.65	2.82	-22.7

Table 17.1 presents the differences in the percentages of fatty acids present in the subcutaneous fat of the two cohorts.

Saturated fatty acids in the adipose tissue increased by 24.5% from 1962 to 1988. The monounsaturated fatty acids decreased by 4%, while the polyunsaturated fatty acids decreased by 30%. As a result of these changes the P:S ratio decreased by 44.6% and the M:S ratio by 22.7%. There is evidence that the low percentage of linoleic acid, which is the main constituent of polyunsaturated fatty acids, in the subcutaneous tissue is correlated closely with the risk of coronary heart disease (62).

The 30% decrease of polyunsaturated fatty acids in the adipose tissue of Cretan males corresponds almost exactly to an analogous (29.3%) increase in death rates from circulatory system diseases in the Greek population from 1960-1964 to 1980-1984 (63). Table 17.2 shows that the Greek population is the only one of 14 developed countries with a rapid increasing death rate due to circulatory system diseases in the past 30 years.

Subcutaneous fat analysis offers strong evidence supporting the dietary changes that have been observed in Greece in the past 30 years. Red meat was consumed on the average, 1.3 times per week in 1962, as compared to 5.4 times in 1988 (60,61). Although the dietary intake of olive oil has not decreased significantly over the past 30 years, there has been a significant reduction in polyunsaturated fatty acids and an analogous increase in saturated fatty acid intake. As a result, the P:S and M:S ratios have decreased significantly in the Cretan population from 1962 to 1988.

Epidemiological studies in other Mediterranean countries support the conclusions reached by the Greek researchers. In southern Italy, Mancini (64) observed an average yearly rise of 2 mg% in plasma cholesterol in conjunction with abandonment of the traditional Mediterranean diet. In Spain, Balaguer-Vintro reported a gradual departure from traditional dietary habits in economically developed regions of Catalonia, along with increasing consumption of saturated animal fats (from meats and dairy products) and a 19% reduction in olive oil consumption. Associated with these changes have been higher plasma cholesterol levels and increased incidence of ischaemia heart disease.

Clinical studies have confirmed olive oil's unique effect on serum cholesterol. Rubba *et al.* (65) and Mancini *et al.* (66) analyzed fatty acid content in adipose tissue, which correlates with dietary fat intake, and concluded that high olive oil intake lowers adipose SFA levels and subsequently reduces the risk of ischemia heart disease. Furthermore, high oleic acid content in adipose tissue corresponded with a low per-

TABLE 17.2

Trends in Mortality from Selected Causes of Death per 100,000 Population and Selected Countries.

Country		Malignancies (8-14)	Diseases of circulatory system (25-30)	Ischemic Heart Disease (27)	Celebrovascular Diseases (29)
Greece	Males 1960-64	118.6	189.0	72.1	67.8
	1980-84	143.5	244.4	82.2	84.3
	Change %	+21	+29	+14	+24
	Females 1960-64	74.9	168.9	53.2	72.9
	1980-84	80.4	185.9	30.0	85.6
	Change %	+7	+10	-43	+17
USA	Males 1960-64	145	503.9	333.6	86.7
	1980-84	164	335.4	204.8	42.2
	Change %	+13	-33	-38	-51
	Females 1960-64	108.5	314.4	168.5	76.3
	1980-84	108.1	189.5	98.1	36.1
	Change %	-0.4	-40	-42	-53
Canada	Males 1960-64	141.9	431.7	296.7	74.6
	1980-84	165.7	299.7	199.7	42.9
	Change %	+17	-31	-33	-43
	Females 1960-64	113.0	277.7	151.6	72.5
	1980-84	108.0	162.8	90.5	35.5
	Change %	-4	-41	-40	-51
Belgium	Males 1960-64	176.6	384.0	116.4	54.8
	1980-84	213.1	312.0	121.9	65.8
	Change %	+20.6	-19	+5	+20
	Females 1960-64	124.9	256.3	54.0	45.5
	1980-84	113.1	186.3	48.7	52.9
	Change %	-9	-27	-10	+16

TABLE 17.2 continued

Trends in Mortality from Selected Causes of Death per 100,000 Population and Selected Countries.

Country		Malignancies (8-14)	Diseases of circulatory system (25-30)	Ischemic Heart Disease (27)	Celebrovascular Diseases (29)
	Males 1960-64	164.3	366.6	219.4	79.2
	1980-84	182.8	326.4	226.2	51.3
Denmark	Change %	+11	-11	+3	-35
	Females 1960-64	134.9	258.3	123.0	74.7
	1980-84	136.3	175.5	101.4	40.1
	Change %	+1	-32	-17.5	-46
	Males 1960-64	192.6	542.4	335.2	120.3
	1980-84	173.7	423.2	273.8	74.2
Finland	Change %	-10	-22	-18	-38
	Females 1960-64	110.0	368.9	158.6	117.3
	1980-84	91.9	212.2	100.0	58.4
	Change %	-14	-42.5	-37	-50
	Males 1960-64	172.5	292.6	73.1	95.8
	1980-84	200.9	219.1	71.0	61.9
France	Change %	+16.5	-25	-3	-35
	Females 1960-64	105.0	184.9	33.4	68.4
	1980-84	91.9	124.1	27.8	42.9
	Change %	-12.5	-33	-17	-37
	Males 1960-64	145.7	365.4	172.0	109.4
	1980-84	188.8	301.6	109.0	80.1
Italy	Change %	+29.5	-17.5	-36.6	-27
	Females 1960-64	100.9	280.2	119.1	84.4
	1980-84	100.0	194.3	48.6	60.4
	Change %	-1	-30.6	-59	-28.5

TABLE 17.2 continued

Trends in Mortality from Selected Causes of Death per 100,000 Population and Selected Countries.

Country		Malignancies (8-14)	Diseases of circulatory system (25-30)	Ischemic Heart Disease (27)	Celebrovascular Diseases (29)
	Males 1960-64	168.1	310.3	181.8	72.1
	1980-84	199.9	287.8	161.2	51.3
Netherlands	Change %	+19	-7	-11	-29
	Females 1960-64	120.7	228.2	101.5	72.2
	1980-84	108.3	151.3	62.7	40.9
	Change %	-10	-34	-38	-43
	Males 1960-64	128.5	281.2	69.6	101.4
	1980-84	153.1	255.7	74.0	80.1
Spain	Change %	+19	-9	+6	-21
	Females 1960-64	89.6	233.3	46.8	87.7
	1980-84	80.7	177.1	29.9	66.6
	Change %	-10	-24	-36	-24
	Males 1960-64	130.8	354.8	218.5	75.4
	1980-84	133.4	333.6	231.8	48.4
Sweden	Change %	+2	-6	+6	-36
	Females 1960-64	110.6	263.7	131.3	72.8
	1980-84	101.8	177.0	99.0	40.2
	Change %	-8	-33	-25	-45
	Males 1960-64	180.6	448.6	264.6	103.2
	1980-84	183.9	363.1	233.6	66.7
U.K.	Change %	+2	-19	-1	-35
	Females 1960-64	115.3	291.0	132.8	90.6
	1980-84	123.8	201.6	96.2	58.2
	Change %	+7	-31	-27.5	-36

TABLE 17.2 continued

Trends in Mortality from Selected Causes of Death per 100,000 Population and Selected Countries.

Country		Malignancies (8-14)	Diseases of circulatory system (25-30)	Ischemic Heart Disease (27)	Celebrovascular Diseases (29)
Australia	Males 1960-64	138.8	501.8	317.4	97.3
	1980-84	164.1	340.0	215.1	64.0
	Change %	+18	-32	-32	-34
	Females 1960-64	98.6	317.0	154.9	95.7
	1980-84	100.4	197.3	98.2	56.7
	Change %	+2	-38	-36.6	-41
Japan	Males 1960-64	138.6	367.8	72.4	235.2
	1980-84	148.2	226.1	40.0	110.5
	Change %	+7	-38.5	-45	-53
	Females 1960-64	96.7	256.2	50.6	159.6
	1980-84	82.7	153.5	22.3	76.6
	Change %	-14.5	-40	-56	-52

centage of polyunsaturated and saturated fatty acids and correlated with dietary fiber. The nutritional value of a MUFA-rich and high-fiber diet was verified in a study conducted by these researchers (67).

After such a diet was given to 200 hyperlipidemic outpatients in a metabolic ward, the serum cholesterol of 100 patients was reduced to moderate levels (58,66,68,69). The researchers emphasized that this result was caused by a decline in LDL-C, while HDL-C was maintained at consistent levels.

Finally, Mancini *et al.* (70) have examined arterial blood pressure in relation to lipid intake. They found that consistent olive oil intake reduces arterial blood pressure. Zoppi *et al.* (71) investigated the effect of olive oil on maintaining a reduction of atherogenous plasma LDL cholesterol. They dispensed a corn oil-based diet to 70 subjects (males and females) with peripheral vascular disease. Half of the subjects were to receive the corn oil diet (with a polyunsaturate/saturate ratio of 1.3) for the duration of the study period; the other half were to change to an olive oil-based diet (with a polyunsaturate/saturate ratio of 0.52) halfway through the period. After 6 months the subjects serum lipid levels were tested and compared.

The continuous corn oil group showed a statistically significant drop in total cholesterol and LDL-C, as well as in HDL-C. The olive oil group showed a significant reduction in LDL cholesterol but, more important, a pronounced rise in HDL cholesterol.

Viola *et al.* (72) studied 32 subjects who received either an olive oil, or a sunflower oil-enriched diet of 80 g supplemented with 20 g of invisible fat. The sunflower oil-based diet resulted in a perceptible drop in LDL cholesterol and apoproteins A and B as well as an inappreciable decrease in HDL cholesterol; the group receiving an olive oil-enriched diet registered a significant drop in LDL-C and an insignificant decrease in apo B but a pronounced increase in HDL cholesterol and in apo A. In a study comparing the effects of saturated, monounsaturated, and polyunsaturated fatty acids on individual lipoprotein fractions, Mattson and Grundy (73) examined 20 men, in a metabolic ward: 8 hypertriglyceridemic patients and 12 patients with normal triglyceride levels. In random order they were administered diets high in fat (40% total calories) either comprising palm oil (high saturated), sunflower oil (high polyunsaturated) or high sunflower oil (high monounsaturated). After 4 weeks, a pronounced drop in LDL cholesterol was noted equally in persons receiving the monounsaturated and the polyunsaturated diets among the normotriglyceridemic patients. Although HDL was unchanged with the monoene diet, it was decreased by the polyunsaturated diet. Thus, although the polyene diet lowered total cholesterol to a greater extent than did the monounsaturated diet, this reduction was due partly to the reduction of HDL cholesterol. In the hypertriglyceridemic patients a 10% reduction in total cholesterol was observed in the monounsaturated and polyunsaturated diets; HDL levels remained unchanged with both diets.

In a study similar to Mattson and Grundy's (73), Becker *et al.* (74) studied the effects of various high-fat diets (40% total calories) on individual lipoprotein fractions. For 4 weeks, 12 subjects were administered in random order diets high in saturated fatty acids, high in monounsaturated fatty acids, and high in polyunsaturated fatty acid. When the saturated diet was replaced by either of the unsaturated diets, the HDL cholesterol levels rose substantially while the LDL cholesterol dropped appreciably. The differences were not as significant as those in Mattson and Grundy's study because the original mean plasma cholesterol levels were lower than average due to the younger age of the subjects.

Mensink and Katan (42) conducted an experimental study comparing the effects on serum lipids of a diet high in complex carbohydrates. Healthy normolipidemic men and women (N=48) received a Western-type diet for 17 days. These free-living subjects were divided then into two groups and were fed conventional mixed solid foods. The first group consumed a low-fat, high fiber, high carbohydrate diet, the second received a monounsaturated diet rich in olive oil. After 36 days, the participant's serum lipid parameters were analyzed and compared. Total mean serum cholesterol dropped 0.44 mmol/l among subjects on the high-carbohydrate diet, versus 0.46 mmol/l among those on the high monounsaturate diet; serum triglycerides increased 0.19 mmol/l with the carbohydrate diet in comparison with a decrease of 0.06 with the olive oil diet. HDL cholesterol levels were reduced 0.19 mmol/l in the carbohydrate diet, despite a high fiber intake whereas they increased 0.83 mmol/l with the high-monounsaturate diet. These results were confirmed by a similar study conducted by Grundy (43). He (43) examined 11 patients in a metabolic ward who received liquid formula diets. The patients randomly received three different diets: a diet high in monounsaturated fat (40% fat, 43% carbohydrates) a diet high in carbohydrate and low in fat (20% fat, 63% carbohydrate) and a diet high in saturated fat (40% fat, 43% carbohydrate). Each diet was administered for 4 weeks. With regard to mean plasma total cholesterol, the monounsaturate-rich diet caused a 13% reduction and the high-carbohydrate diet a reduction of 8% in comparison to the diet high in saturated fat. Similarly the monounsaturate-rich diet lowered LDL cholesterol by 21% and the high-carbohydrate by 15%. Furthermore, the high-carbohydrate diet, raised triglyceride levels perceptibly and resulted in an appreciable drop in HDL cholesterol, while the monounsaturated diet had a neutral effect on these parameters. Grundy (43) concluded that a high-fat diet containing primarily monounsaturated fatty acids was as effective as low-fat, high-carbohydrate diet, but at the same time uniquely raised the HDL/LDL cholesterol ratio.

Baggio *et al.* (75) studied 11 young volunteers who were administered an olive oil-enriched diet in comparison with a low-fat diet in a metabolic ward. Serum lipoprotein levels biliary cholesterol saturation index, and gall bladder motility were analyzed at the end of the study. It was reported that mean total cholesterol was reduced by an appreciable 9.5%, total apo B 7.4%, LDL-C 12.2% and total triglycerides 25.5%, while HDL cholesterol and apo A mean levels were unaltered. They (75) concluded that olive oil appeared to be a unique nutritional source of dietary lipids applicable to the treatment of mild hypercholesterolemia. Its favorable effect on HDL and on apo A serum levels make it as therapeutically useful as PUFAs in lowering cholesterol levels.

Sirtori *et al.* (76) reported an experimental study comparing the effect of low-fat diets with that of olive oil or corn oil enriched diets, on serum lipids and platelets. Twenty-three middle-aged subjects having risk factors for coronary heart disease were examined. For 8 weeks the patients received either an olive oil-rich dict (with a polyunsaturate/saturate ratio of 0.33) or a corn oil-rich diet (with a PS ratio of 1.28). Whereas total cholesterol was reduced by corn oil, HDL cholesterol and apo A levels were increased by olive oil alone. Plasma apo B was lowered by both olive oil and corn oil diets, but the apo A: apo B ratio was increased only by olive oil.

The conclusion reached from the epidemiological and clinical studies is that monounsaturated oleic acid, found in olive oil is as effective as PUFAs in lowering total serum cholesterol, LDL cholesterol and triglyceride levels, while at the same time has a positive effect on HDL cholesterol. HDL is a protective factor against coronary heart disease. The unique effect of olive oil (rich in oleic) on serum lipids, its favorable ratio between monounsaturated and saturated fatty acids and between PUFAs and antioxidants (Vitamin E), and its high content of essential fatty acids give it high biological and nutritional value and make it an excellent source of dietary lipids. This dietary strategy, an important facet of the Mediterranean diet, provides successful protection against the onset of atherosclerosis by controlling plasma cholesterol adequately without any harmful side effects.

Conclusion

The increasing availability of food and the decrease in manual work and strenuous activity that accompany technological progress are resulting in striking changes in dietary patterns and eating habits. The currently evolving popular diet is high in dietary fat (saturated and polyunsaturated), high in cholesterol and animal protein, and low in fiber and complex carbohydrates. Many people's diets are characterized by energy excess, deficiency in vitamins, minerals, and fiber and an unbalanced nutritional intake which results in an increased risk of chronic diseases, such as CHD, diabetes, and cancer. The nutritional solution for many people would be the adoption of a sound, healthy diet similar to the Mediterranean diet. This diet is wholesome and nourishing. It entails a high intake of complex carbohydrates (whole wheat bread, legumes, and wheat products), fresh fruit and vegetables and fish (high in ω-3 FAs), low consumption of dairy products and high olive oil intake. Olive oil as a lipid source has a unique effect on serum lipids. In addition olive oil seems to have effective cholagogic properties and a therapeutical effect on peptic ulcers. This oil, rich in monounsaturated fatty acids, also has positive effects on growth and development and seems to have beneficial effects on aging. Indeed olive oil's nutritional value is reflected in the health of its consumers in the Mediterranean basin: a low rate of CHD and a high life expectancy. These benefits, however, are progressively being abolished by the Westernization of the traditional diet.

References

1. Viola, P. and M. Audisio, Olive oil and health. International Olive Oil Council, Madrid, Spain (1987).
2. Dianzani, M.U., La persossidazione lipidica nella patogenesi delle lesioni cellulari. In Simposio su "Il contenuto ottimale di acido linoleico nella dieta", Sestri Levante (Italia) (1978).
3. Babior, M.E., Oxygen-dependent microbia killing of phagocytes, N. Engl. J. of Med. 298:659 (1977).
4. Graham, S., J. Marshall, C. Mettlin, T. Rzepka, T. Nemoto, Diet in the epidemiology of breast cancer, Am. J. Epidem. 116:65 (1982).
5. Vitale, J.J., S.A. Broitman, Lipids and Immune Function Cancer Res. 41:3706 (1981).

6. Santamaria, L., A. I. Bianchi, caroteniodi e le vitamine anti-ossidanti nella prevenzione alimentare di malattie neoplastiche e dell'invecchiamento. In Simposio su Rilettura di un problema, i lipidi alimetari Rimini (Italia) (1984).
7. Harman, D., Free radical theory of aging: Effect of fat on lipid composition and function on the brain. In: 3rd International Congress on the Biological Value of Olive Oil, Chanea, Crete, Greece (1980).
8. Timiras, P.S., Radicali liberi, malattie ed invecchiamento, Fed. Med. 8:890 (1982).
9. Toffano, G. and G. Calderini, Nuove prospettive sperimentali dell invecchiamento cerebrale. Fidia Biomed. Inform Anno 1, n. 2 aprile maggio (1984).
10. Turchetto, E. and C. Pignatti, Protective role of vitamin E on essential fatty acids. Acta Vitaminol. Enzymol, 4:215 (1982).
11. Keys, A. Mortalita in 15 anni nelle coorti del Seven Countries Study. In 1st Congr. Nazionale di Terapia, Roma (Italia) (1985).
12. Ewald, C.A. and J. Boas, Virchows Archives of Pathological Anatomy 104:271 (1889).
13. Castelli, W.P., G.T. Doyle, and T. Gordon, HDL cholesterol and other lipids in coronary heart disease: The cooperative lipoprotein phenotyping study. Circulation 55:767 (1977).
14. Benvestito, F., T. Boscia, R. Moschetta, Olio di oliva e funzionamento gastrico In: 1st International Congress in the Biological Value of Olive Oil, Lucca, Italy (1969).
15. Taits, N.S., Use of olive oil in the treatment of ulcer patients. Urach. Del. 7:67, (1966).
16. Charbonnier, A., Recenti acquisizioni sul valore biologico dell'olio di oliva in Francia. In 1st Congr. Nazionale di Terapia, Roma (Italia) (1985).
17. Ricci, G., L'olio di oliva in gastroenterologia. In 1st International Congress on the Biological Value of Olive Oil, Lucca, Italy (1969).
18. Demole, M., Cholagogues et cholecystokinetiques alimetaires dans le traitement des dyskinesies billiaires. In 1st International Congress on the Biological Value of Olive Oil, Lucca, Italy (1969).
19. Pavel, I., La vesicule biliaire et ses voies d'excretion. Masson Edit. Paris 1:228 (1959).
20. Heaton, K.W., Epidemiologia della calcolosi biliare colesterolica. Ruolo della dieta. In 3rd Simposio sulla fisiologia e trattamento della calcolosi biliare colesterolica con acido chenodesossicolico, Cortina d' Ampezzo (Italia) (1977).
21. Sarles, H., J. Hanton, N.E. Planche, H. Lafont and A. Gerolami, Diet cholesterol gallstones and composition of bile. Am. J. Dig. Dis. 15:251 (1970).
22. Den Resten, L., W.E. Connor and S. Bell, The effects of dietary cholesterol on the composition of human bile. Surgery 73:266 (1973).
23. Nervi, F. and J.M. Dietschy, Ability of six different lipoproteic fractions to regulate the rate of hepatic cholesterogenesis in vivo. Bioch. Biophys. Acta

486:387 (1977).
24. Coronary drug project group. Gallbladder suffering in cholesterol-lowering treatment. N. Engl. J. Med 296:1186 (1977).
25. Barbara, L., F. Bazzoli and E. Roda, La calcolosi biliare colesterolica: patogenesi e terapia. Fed. Med. 36:644 (1983).
26. Grundy, S.M., Effect of polyunsaturated fats on lipid metabolism in patients with hypertriglyceridemia. J. Clin. Invest. 55:288 (1975).
27. Moore, R.B., J.T. Anderson, H.L. Taylor, A. Keys and D. Frantz, Effect of dietary fat in fecal excretion of cholesterol and its degradation products in man. J. Clin. Invest. 47:1517 (1968).
28. Bucko, A. and E. Ginter, Les possibilities de l'utilisation therapeutic de l'huile d'olive dans la gastroenterologie. In 2nd International Congress on the Biological Value of Olive Oil, Torremolinos (Spain) (1975).
29. Borgman, R.F. and F.H. Haseldem, Cholelithiasis in rabbits. Effect of diet upon formation and dissolution of gallstones. Am. J. Vet. Res. 29:1287 (1968).
30. Schuller, A., M.D. Rubio, J. Esteban and L. Erroz, Variaciones de la composicion de la bilis en relacion con la grasa de la dieta. In 3rd International Congress on the Biological Value of Olive Oil, Chanea, Crete, Greece (1980).
31. Messini, M. and M. Cairella, Dietetica, Ed. SEU, Roma (1967).
32. Crawford, M.A., The role of essential fats in human maternal and infant nutrition with special reference to brain growth. In 2nd International Congress on the Biological Value of Olive Oil, Torremolinos (Spain) 6-9 p. 35, May (1975).
33. Galli, C., Comparative effects of olive oil and other edible fats on brain development. In 12th World Congr. of Intern. Soc. for Fat Res. Milan (Italy) Abstr. 115, 2-7 Sept. (1974).
34. Laval-Jeantent, A.M., M. Laval-Jeantent, C. Bergot and J. Gousaud, Effects des lipides oleiques sur la croissance et la composition de l'os en nutrition experimentale. In 3rd International Congress of the Biological Value of Olive Oil. Chanea, Crete, Greece (1985).
35. Harman, D., Effect of dietary polyunsaturated fats on aging and degenerative diseases. In Symposium on "contenuto ottimale de acido linoleico nella dieta". Sestri Levante (Italia) (1978).
36. Manfredi-Romanini, M.G., La senescenza come problema biologico. Fed. Med. 3:253 (1984).
37. Harman, D., Free radical theory of aging: Effect of the degree of unsaturation of dietary fat on mortality rate. J. Geront. 26:451 (1971).
38. Porta, E.A., N.S. loun and R.T. Mita, Effects of the type of dietary fats at two levels of vitamin E in Wistar male rats during development and aging Merch. Aging and Devel. 16:63 (1980).
39. Harman, D., Free radical theory of aging: Effect of fat on lipid composition and function on the Brain. In 3rd International Biological Value of Olive Oil. Chanea, Crete, Greece (1980).

40. Harman, D., Free radical theory of aging: Effect of fat on mortality rate in 1st International Congress on the Biological Value of Olive Oil, Lucca (Italy) (1969).
41. Garg, A., A. Bonanone, S.M. Grundy, Z. Zhang and R.H.S. Unger, Comparison of a high carbohydrate diet with a high monounsaturated fat diet in patients with non-insulin dependent diabetes mellitus. N. Eng. J. Medicine 319:829 (1988).
42. Mensink, R.P., and M.B. Katan, Effect of a diet enriched with monounsaturated or polyunsaturated fatty acids on levels of low-density and high density lipoproteins cholesterol in healthy women and men. New Engl. J. Med. 321:436 (1989).
43. Grundy, S.M., Comparison of monounsaturated fatty acids and carbohydrates for lowering plasma cholesterol. N. Engl. J. Med. 314:745 (1986).
44. Sestoft, L., T. Krarup, B. Palmvig, H. Meinertz and O. Faergeman, High carbohydrate, low-fat diet: Effect on lipid and carbohydrate metabolism. GIP and insulin secretion in diabetics. Dan. Med. Bull. 32:64 (1985).
45. Coulston, A.M., C.B. Hollenbeck, A.L.M. Swislocki, Chen-Y-Di, and G.M. Reaven, Deleterious metabolic effects of high carbohydrate sucrose-containing diets in patients with non-insulin dependent diabetes mellitus. Am. J. Med. 82:213 (1987).
46. Bierman, E.L. and J. T. Hamlin, The hyperlipemic effect of a low-fat, high carbohydrate diet in diabetic subjects. Diabetes 10:432 (1961).
47. Anderson, J.W., Effect of carbohydrate restriction and high carbohydrate diets on men with chemical diabetes. Am. J. Clin. Nutr. 30:402 (1977).
48. Glauber, H, P. Wallace, K. Griver and G. Brechtel, Adverse metabolic effect of omega-3 fatty acids in non-insulin dependent mellitus. Ann. Intern. Med. 108:633 (1988).
49. Keys, A., Coronary heart disease in Seven Countries Study. Circulation 44 Suppl.1, (1970).
50. Ahrens, E.H., J. Hirsch and W. Insull, The influence of dietary fats on serum lipid levels in man. Lancet. 1:943 (1975).
51. Keys, A., Coronary Heart Disease. The Global Picture-Atherosclerosis, 22:149 (1975).
52. Christsakis, G., M.K. Fordyce and C.S. Kurtz, The biological and medical aspects of olive oil. Inter. Olive Oil Council, Madrid, Spain (1982).
53. Harvengt, C., High density lipoproteins and atherosclerosis. Acta Clin. Belg. 33:73 (1978).
54. Castelli, W.P., G.T. Doyle and T. Gordon, HDL cholesterol and other lipids in coronary heart disease: The cooperative lipoprotein phenotyping study. Circulation 55:767 (1977).
55. Shepherd, J., C.J. Packard, J.R. Patsch, A.M. Gotto, Jr., and O.D. Tauton, Effects of dietary polyunsaturated fat on the properties of high density lipoproteins and the metabolism of apolipoprotein A-1-J. Clin. Invest. 61:1582 (1978).

56. Schaefer, E.J., R.I. Levy, N.D. Ernst, F.D. Van Sant and H.B. Brewer, The effects of a low cholesterol, high polyunsaturated fat, and low fat diets on plasma lipid and lipoprotein cholesterol level in normal and hypercholesterolemic subjects. Am. J. Clin. Nutr. 34:1758 (1981).
57. Ferro-Luzzi, A., P. Strazzulo, C. Scaccini, A. Siana, S. Sette, M.A. Mariani, P. Mastranzo, R.M. Dougherty, J.M. Iacono and M. Mancini, Changing the Mediterranean diet: Effects on blood lipids. Am. J. Clin. Nutr. 40:1027 (1984).
58. Mancini, M., E. Farinaro, A. Postiglione, P. Rubba and O. Vaccaro, Eating Neapolitan style: a possible treatment of hyperlipidemia. In Recent advances in clinical nutrition. A.H. Howard, I. McLean Baird Eds., J. Libbey London, (1981).
59. Aravanis, C., The Greek islands heart study. In 3rd International Congress on the Bilogical Value of Olive Oil. Chanea, Crete, Greece (1980).
60. Kafatos, A., I. Kouroumallis, N. Facourelis and J. Vlachonikolis, Adipose tissue fatty acids composition of obese and non-obese adult Cretans. The 14th International Congress of Nutrition. Seoul, Korea (1989).
61. Christakis, G., E.L. Severinghaus, Z. Maldonado, F. Kafatos, and S.A. Hashim, Crete: A study in the metabolic epidemiology of coronary heart disease. The Am. J. of Cardiology 15:320 (1965).
62. Beynen, A.C. and M.B. Katan, Rapid sampling and long-term storage of subcutaneous adipose tissue biopsies for determination of fatty acid composition. The Am. J. Clin. Nutrition 42:317 (1985).
63. WHO annual statistical yearbook (1988).
64. Mancini, M., Patogenesi dell aterosclerosi. In 1st Congr. Naz. Terapia, Roma, (Italia) (1985).
65. Rubba, P., M. Mancini, F. Fidanza, G. Gautiero, T. Nikkari and M.F. Oliver, Adipose tissue fatty acids and blood pressure in middle aged men from southern Italy. Proceedings 2nd. Int. Congr. on Essential Fatty Acids, Prostaglandins and Leukotrienes, Ads., London (1985).
66. Mancini, M., A. Postiglione, P. Rubba and E. Farinaro, Concentration plasmatica de las lipoproteinas de alta densidad en pacientes hiperlipidemicos sometidos a dieta reductora de lipidos, en regimen ambulatorio. In Aspectos Actuales de las Hiperlipoproteinemias, R. Carmena Ed. p. 179, Editorial Garsi, Madrid (1981).
67. Mancini, M. G. Riccardi and A. Rivellese, Prevention de l'atherosclerose par lipides et fibres alimentaires. Med. Chir. Dig. 11:601 (1982).
68. Mancini, M., A. Postiglione, P. Rubba and E. Farinaro, Trattamento dietetico delle iperlipidemie: analogie con la dieta mediterranea. In: 3rd International Congress on the Biological Value of Olive Oil, Chanea, Crete, Greece (1980).
69. Farinaro, E., P. Rubba, G. Postiglione, G. Ricardi and M. Mancini, Studies on diet and plasma lipids in Naples. Progr. Biochem. Pharmacol. 19:123 (1983).

70. Mancini, M., P. Rubba and P. Strazzullo, Consumo di olio di oliva e regolazione della pressione arteriosa. In 1 Congr. Naz. Terapia, Roma, (Italia) (1985).
71. Zoppi, S., C. Vergani, P. Giorgetti, S. Rapelli and B. Berra, Effectiveness and reliability of medium term treatment with a diet rich in olive oil of patients with vascular diseases. Acta Vitaminol. Enzymol. 7:3 (1985).
72. Viola, P., V. Tonelli and M. Audisio, Olio di olive e microcircolo. In 1st Congr. Naz. Terapia, Roma (Italia) (1985).
73. Mattson, F.H. and S.M. Grundy, Comparison of effects of dietary saturated, monounsaturated, and polyunsaturated fatty acids on plasma lipids and lipoproteins. J. Lipid Res. 26:194 (1985).
74. Becker, N., D.R. Illingworth, P. Alaupovic, W.E. Connor and E.E. Sundberg, Effects of saturated, monounsaturated, and omega-6 polyunsaturated fatty acids on plasma lipids, lipoproteins and apoproteins in humans. Am. J. Clin. Nutr. 37:355 (1983).
75. Baggio, G., A. Pagnan, M. Muraca, S. Martini, A, Opportuno, A. Bonanone, G.B. Ambrosio, S. Ferrari, P. Guarini, D. Piccolo, E. Manzato, R. Correcher and G. Crepaldi, Olive Oil enriched diet: effect on serum lipoprotein levels and biliary cholesterol saturation. Am. J. Clin. Nutr. 47:960 (1988).
76. Sirtori, C.R., E. Tremoli, E. Gatti, G. Montanari, M. Sirtori, S. Colli, G. Gianfranceschi, P. Maderna, C.Z. Dentone and G. Testolin, Controlled evaluation of fat intake in the Mediterranean diet: Comparative activities of olive oil and corn oil on plasma lipids and platelets in high-risk patients. Am. J. Clin. Nutr. 44:635 (1986).

Epilogue

Future Research Needs

A multilevel research approach is recommended for improvement in quality and increase in quantity of olive oil. Starting at the production level, the search for olive tree cultivars best suited for each cultivation area should be continued. It is deemed desirable to consider the newer genetic engineering techniques in order to reach better health and productivity states for the trees. The search for improved ways to neutralize, and eventually eradicate, the traditional enemy of olive cultivation, the olive fly *Dacus oleae*, should be intensified.

The olive fruit harvesting methods must be modernized. As farm labor becomes increasingly scarce, the mechanical means of collecting the fruit properly and efficiently must be perfected. Improved ways of storing the fruit before processing should be also sought. The need for better fruit storage facilities is greater for olive mills of limited capacity. Although significant progress has been made in designing efficient olive processing equipment, there is room for improving the rate of oil extraction from the fruit. The use of enzymes facilitating the release of oil from the tissue should be further explored.

As natural antioxidants hold promise for prolonging the market life of olive oil, additional work on the application of these additives to the preservation of olive oil is indicated. The contention of a particularly beneficial effect of olive oil consumption on human health needs additional justification. The trans isomerization of oleic acid on heating olive oil for culinary purposes invites investigation.

Finally, the utilization of by-products of the olive oil industry should be further explored. An example of a potentially useful olive by-product is the anthocyanin present in the ripe olive skin. Another example is the sugars which are present in the aqueous phase of the olive tissue and which could be subjected to alcoholic fermentation after the extraction of the oil.

Index

A

B

E

F

G

H

I

U

V

W

Y